Mathematical Summary for Digital Signal Processing Applications with Matlab

E.S. Gopi

Mathematical Summary for Digital Signal Processing Applications with Matlab

Springer

E.S. Gopi
National Institute of Technology, Trichy
Dept. Electronics & Communication Engineering
67 Tanjore Main Road
Tiruchirappalli-620015
National Highway
India
esgopi@nitt.edu

ISBN 978-90-481-3746-6 e-ISBN 978-90-481-3747-3
DOI 10.1007/978-90-481-3747-3
Springer Dordrecht Heidelberg London New York

Library of Congress Control Number: 2009944069

Cover design: eStudio Calamar S.L.

Printed on acid-free paper

Springer is part of Springer Science+Business Media (www.springer.com)

Dedicated to my son G.V. Vasig and my wife G. Viji

Preface

The book titled "Mathematical summary for Digital Signal Processing Applications with Matlab" consists of Mathematics which is not usually dealt in the DSP core subject, but used in the DSP applications. Matlab Illustrations for the selective topics such as generation of Multivariate Gaussian distributed sample outcomes, Optimization using Bacterial Foraging etc. are given exclusively as the separate chapter for better understanding. The book is written in such a way that it is suitable for Non-mathematical readers and is very much suitable for the beginners who are doing research in Digital Signal Processing.

E.S. Gopi

Acknowledgements

I am extremely happy to express my thanks to Prof. K.M.M. Prabhu, Indian Institute of Technology Madras for his constant encouragement. I also thank the Director Prof. M. Chidambaram, Prof. B. Venkataramani and Prof. S. Raghavan National Institute of Technology Trichy for their support. I thank those who directly or indirectly involved in bringing up this Book. Special thanks to my parents Mr. E. Sankarasubbu and Mrs. E.S. Meena.

Contents

Chapter 1
Matrices

One-dimensional array representation of scalars are called vector. If the elements are arranged in row wise, it is called Row vector. In the same fashion, if the elements of the vector are arranged in column wise, it is called column vector. Two-dimensional array representations of scalars are called matrix. Size of the matrix is represented as R×C, where R is the number of rows and C is the number of columns of the matrix. Scalar elements in the array can be either complex numbers ($\mathbb{C}$) or the real numbers. ($\mathbb{R}$). The column vector is represented as $\underline{X}$. The Row vector is represented as $\underline{X}^T$.

Example 1.1. Row Vector with the elements filled up with real numbers

$$[2.89 \quad 21.87 \quad 100]$$

Column Vector with the elements filled up with Complex numbers.

$$\begin{bmatrix} 1+j \\ -j \\ 9+7j \\ 0 \end{bmatrix}$$

Matrix of size 2×3 with the elements filled up with real numbers

$$\begin{bmatrix} 2 & 3 & 6 \\ 4 & 1 & 2 \end{bmatrix}$$

Matrix of size 3×2 with the elements filled up with complex numbers

$$\begin{bmatrix} -j & 1+j \\ -2j & 5j \\ 0 & j \end{bmatrix}$$

E.S. Gopi, *Mathematical Summary for Digital Signal Processing Applications with Matlab*, DOI 10.1007/978-90-481-3747-3_1,

1.1 Properties of Vectors

Scalar multiplication of the vector $\underline{X}$ is given as c $\underline{X}$, where c $\in \mathbb{R}$.

Example 1.2.

$$2 * \begin{bmatrix} 1+j \\ -j \\ 9+7j \\ 0 \end{bmatrix} = \begin{bmatrix} 2+2j \\ -2j \\ 18+14j \\ 0 \end{bmatrix}$$

Linear combinations of two vectors $\underline{X_1}$ and $\underline{X_2}$ are obtained as $c1^*\underline{X_1} + c2^*\underline{X_2}$, where c1, c2 $\in \mathbb{R}$

Example 1.3.

$$2 * \begin{bmatrix} 1+j \\ -j \\ 9+7j \\ 0 \end{bmatrix} + 3 * \begin{bmatrix} 1-j \\ j \\ -7j \\ 0 \end{bmatrix} = \begin{bmatrix} 5-j \\ j \\ 18-7j \\ 0 \end{bmatrix}$$

Example 1.4. Graphical illustration of summation of two vectors [3 1] and [1 2] to obtain the vector [4, 3] is given below (Fig. 1.1). (Recall the Parallelogram Rule of addition).

Note that the first and second elements of the vector are represented as the variable X1 and X2, respectively. The variables X1 and X2 can be viewed as the random variables.

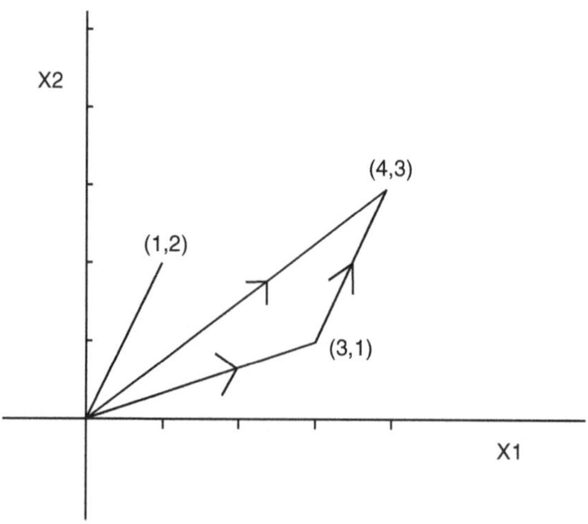

Fig. 1.1 Graphical illustration of the summation of two vectors

1.2 Properties of Matrices

(a) Matrix addition

Let the Matrix A be represented as

$$\begin{bmatrix} a_{11} & a_{12} & a_{13} & \ldots & a_{1m} \\ a_{21} & a_{22} & a_{33} & \ldots & a_{2m} \\ a_{31} & a_{32} & a_{33} & \ldots & a_{3m} \\ . & . & . & . & . \\ a_{n1} & a_{n2} & a_{n3} & \ldots & a_{nm} \end{bmatrix}$$

Also let the Matrix B be represented as

$$\begin{bmatrix} b_{11} & b_{12} & b_{13} & \ldots & b_{1m} \\ b_{21} & b_{22} & b_{23} & \ldots & b_{2m} \\ b_{31} & b_{32} & b_{33} & \ldots & b_{3m} \\ . & . & . & . & . \\ b_{n1} & b_{n2} & b_{n3} & . & b_{nm} \end{bmatrix}$$

$$\Rightarrow \mathrm{A} + \mathrm{B} = \begin{bmatrix} a_{11} + b_{11} & a_{12} + b_{12} & a_{13} + b_{13} & \ldots & a_{1m} + b_{1m} \\ a_{21} + b_{21} & a_{22} + b_{22} & a_{23} + b_{23} & \ldots & a_{2m} + b_{2m} \\ a_{31} + b_{31} & a_{32} + b_{32} & a_{33} + b_{33} & \ldots & a_{3m} + b_{3m} \\ . & . & . & . & . \\ a_{nm} + b_{nm} & a_{nm} + b_{nm} & a_{nm} + b_{nm} & \ldots & b_{nm} + b_{nm} \end{bmatrix}$$

Note that **(i,j)th** element of the matrix 'A' is represented as a_{ij}. Matrix 'A' in general is represented as $\mathrm{A} = [a_{ij}]$

Let $\mathrm{C} = \mathrm{A} + \mathrm{B} \Rightarrow [c_{ij}] = [a_{ij}] + [b_{ij}]$

(b) Scalar multiplication

Let $c \in \mathbb{C}$ or $c \in \mathbb{R}$

$$\Rightarrow \mathrm{cA} = \mathrm{c}[a_{ij}] = [ca_{ij}]$$

(c) Matrix multiplication

The product of the matrix 'A' with size n × p and the matrix 'B' with size p × m is the matrix C of size n × m. The elements of the matrix C are obtained as follows.

$$\mathrm{C} = \begin{matrix} c_{11} & c_{12} & c_{13} & \ldots & c_{1m} \\ c_{21} & c_{22} & c_{23} & \ldots & c_{2m} \\ c_{31} & c_{32} & c_{33} & \ldots & c_{3m} \\ . & . & . & . & . \\ c_{n1} & c_{n2} & c_{n3} & \ldots & c_{nm} \end{matrix}$$

Where $c_{ij} = a_{i1}b_{1j} + a_{i2}b_{2j} + a_{i3}b_{3j} + \ldots + a_{ip}b_{pj}$

Example 1.5. Consider the matrix A and B of size 2 × 3 and 3 × 3, respectively

$$A = \begin{bmatrix} a_{11} & a_{12} & a_{13} \\ a_{21} & a_{22} & a_{23} \end{bmatrix}$$

$$B = \begin{bmatrix} b_{11} & b_{12} & b_{13} \\ b_{21} & b_{22} & b_{23} \\ b_{31} & b_{32} & b_{33} \end{bmatrix}$$

Let the matrix C = AB

$$= \begin{bmatrix} a_{11}b_{11} + a_{12}b_{21} + a_{13}b_{31} & a_{11}b_{12} + a_{12}b_{22} + a_{13}b_{32} & a_{11}b_{13} + a_{12}b_{23} + a_{13}b_{33} \\ a_{21}b_{11} + a_{22}b_{21} + a_{23}b_{31} & a_{21}b_{12} + a_{22}b_{22} + a_{23}b_{32} & a_{21}b_{13} + a_{22}b_{23} + a_{23}b_{33} \end{bmatrix}$$

The matrix A can be viewed as $A = [\underline{a_1}\ \underline{a_2}\ \underline{a_3}]$, where

$$\underline{a_1} = \begin{bmatrix} a_{11} \\ a_{21} \end{bmatrix} \quad \underline{a_2} = \begin{bmatrix} a_{12} \\ a_{22} \end{bmatrix} \quad \underline{a_3} = \begin{bmatrix} a_{13} \\ a_{23} \end{bmatrix}$$

Similarly the matrix B can be viewed as $B = \begin{bmatrix} \underline{b_1} \\ \underline{b_2} \\ \underline{b_3} \end{bmatrix}$, where

$$\underline{b_1} = [b_{11} \quad b_{12} \quad b_{13}]$$
$$\underline{b_2} = [b_{21} \quad b_{22} \quad b_{23}]$$
$$\underline{b_3} = [b_{31} \quad b_{32} \quad b_{33}]$$

So the matrix C = AB can also be represented as $\underline{a_1}\ \underline{b_1} + \underline{a_2}\ \underline{b_2} + \underline{a_3}\ \underline{b_3}$
Also the matrix AB can be obtained as

$$\left[b_{11}(\underline{a_1}) + b_{21}(\underline{a_2}) + b_{31}(\underline{a_3}) \ \ b_{12}(\underline{a_1}) + b_{22}(\underline{a_2}) + b_{32}(\underline{a_3}) \ \ b_{13}(\underline{a_1}) + b_{23}(\underline{a_2}) + b_{33}(\underline{a_3})\right]$$

(d) Matrix multiplication is associative
Let A and B be the two matrices, then A (BC) = (AB) C.

(e) Matrix multiplication is non-commutative
Let A and B be the two matrices, then AB ≠ BA

(f) Block multiplication
Consider the matrix P as shown below

$$\begin{bmatrix} a_{11} & a_{12} & a_{13} & b_{11} & b_{12} \\ a_{21} & a_{22} & a_{23} & b_{21} & b_{22} \\ c_{11} & c_{12} & c_{13} & d_{11} & d_{12} \\ c_{21} & c_{22} & c_{23} & d_{21} & d_{22} \\ c_{31} & c_{32} & c_{33} & d_{31} & d_{32} \end{bmatrix}$$

The above matrix 'P' can be viewed as the matrix with each element filed up with matrix as mentioned below. This way of representing the matrix is called as Block matrix.

$$\begin{bmatrix} \begin{bmatrix} a_{11} & a_{12} & a_{13} \\ a_{21} & a_{22} & a_{23} \end{bmatrix} & \begin{bmatrix} b_{11} & b_{12} \\ b_{21} & b_{22} \end{bmatrix} \\ \begin{bmatrix} c_{11} & c_{12} & c_{13} \\ c_{21} & c_{22} & c_{33} \\ c_{31} & c_{32} & c_{33} \end{bmatrix} & \begin{bmatrix} d_{11} & d_{12} \\ d_{21} & d_{22} \\ d_{31} & d_{32} \end{bmatrix} \end{bmatrix}$$

In short the matrix P is represented as

$$\begin{bmatrix} [A] & [B] \\ [C] & [D] \end{bmatrix}$$

Similarly consider the matrix Q represented as

$$\begin{bmatrix} [E] \\ [F] \end{bmatrix}$$

Then the matrix PQ is represented as

$$\begin{bmatrix} [AE + BF] \\ [CE + DF] \end{bmatrix}$$

This way of multiplying two Block matrices is called as Block multiplication.

(g) Transpose of the matrix

The transpose of the matrix A is the matrix B whose elements are related as follows.

$$a_{ij} = b_{ji}$$

Note that transpose of the Block matrix

$$\begin{bmatrix} [A] & [B] \\ [C] & [D] \end{bmatrix}$$

is given as

$$\begin{bmatrix} [A^T] & [C^T] \\ [B^T] & [D^T] \end{bmatrix}$$

(h) Square matrix

The matrix with number of Rows is equal to the number of Columns.

(i) Identity matrix

The square matrix with all the elements is filled up with zeros except the diagonal elements which are filled up with all ones.

(j) Lower triangular matrix

The square matrix with all the elements is filled up with zeros except the elements in the diagonal and below the diagonal which are filled up at least one non-zero elements.

In other words, Lower triangular matrix is the matrix with zeros in the upper triangular portion of the matrix with at least one non-zero element in the remaining portion.

(k) Upper triangular matrix

The square matrix with all the elements is filled up with zeros except the elements in the diagonal and above the diagonal which are filled up at least one non-zero elements.

In other words, Upper triangular matrix is the matrix with zeros in the Lower triangular portion of the matrix with at least one non-zero element in the remaining portion.

(l) Diagonal matrix

The square matrix with all the elements is filled up with zeros except the diagonal elements which are filled up with at least one non-zero element in the diagonal

(m) Permutation matrix

Permutation matrix is one when multiplied with the matrix interchanges the elements of the matrix column wise or row wise. If the matrix is multiplied by the permutation matrix, columnwise interchange of the elements of the matrix occurs. Similarly if the permutation matrix is multiplied by some matrix, row wise interchange of the elements of the matrix occurs.

Example 1.6. Arbitrary matrix $A = \begin{bmatrix} 1 & 2 & 3 \\ 4 & 5 & 6 \\ 7 & 8 & 9 \end{bmatrix}$

$$\text{Arbitrary Permutation matrix } P = \begin{bmatrix} 1 & 0 & 0 \\ 0 & 0 & 1 \\ 0 & 1 & 0 \end{bmatrix}$$

$$A * P = \begin{bmatrix} 1 & 3 & 2 \\ 4 & 6 & 5 \\ 7 & 9 & 8 \end{bmatrix}$$

Note that the elements of the second column and third column is interchanged using the operation A*P.

$$P * A = \begin{bmatrix} 1 & 2 & 3 \\ 7 & 8 & 9 \\ 4 & 5 & 6 \end{bmatrix}$$

Note that the elements of the second row and third row is interchanged using the operation P*A.

P*P* ... P is always some permutation matrix. Also note that the identity matrix is the trivial permutation matrix, which when multiplied with any arbitrary matrix will end up with the same matrix. Also note that the inverse of any arbitrary permutation matrix is always the permutation matrix.

(n) Inverse of the matrix
Inverse of the matrix is defined only for the square matrix. The matrix A is defined as the inverse of the matrix B if AB = BA = I, where 'I' is the identity matrix. If there exists the inverse matrix for the particular square matrix A, then that matrix 'A' is known as the Invertible matrix. Otherwise it is called as the non-invertible matrix.

1.3 LDU Decomposition of the Matrix

The matrix A as shown below can be represented as the product of three matrices Lower triangular matrix (L) with all ones in the diagonal elements, Diagonal matrix (D) and the upper triangular matrix (U) with all ones in the diagonal elements.

Example 1.7. LDU Decomposition of the matrix $A = \begin{bmatrix} 1 & 2 & 3 \\ 2 & 3 & 4 \\ 3 & 4 & 6 \end{bmatrix}$

$$\Rightarrow \begin{bmatrix} 1 & 0 & 0 \\ 0 & 1 & 0 \\ 0 & 0 & 1 \end{bmatrix} \begin{bmatrix} 1 & 2 & 3 \\ 2 & 3 & 4 \\ 3 & 4 & 6 \end{bmatrix} = \begin{bmatrix} 1 & 2 & 3 \\ 2 & 3 & 4 \\ 3 & 4 & 6 \end{bmatrix}$$

Note: Row operation on the Identity matrix in the LHS and the same operation done on the RHS will not affect the equality.

R2- > R2-2*R1

R3- > R3-3*R1

$$\Rightarrow \begin{bmatrix} 1 & 0 & 0 \\ -2 & 1 & 0 \\ -3 & 0 & 1 \end{bmatrix} \begin{bmatrix} 1 & 2 & 3 \\ 2 & 3 & 4 \\ 3 & 4 & 6 \end{bmatrix} = \begin{bmatrix} 1 & 2 & 3 \\ 0 & -1 & -2 \\ 0 & -2 & -3 \end{bmatrix}$$

$$\Rightarrow \begin{bmatrix} 1 & 0 & 0 \\ 0 & 1 & 0 \\ 0 & 0 & 1 \end{bmatrix} \begin{bmatrix} 1 & 0 & 0 \\ -2 & 1 & 0 \\ -3 & 0 & 1 \end{bmatrix} \begin{bmatrix} 1 & 2 & 3 \\ 2 & 3 & 4 \\ 3 & 4 & 6 \end{bmatrix} = \begin{bmatrix} 1 & 2 & 3 \\ 0 & -1 & -2 \\ 0 & -2 & -3 \end{bmatrix}$$

Again applying Row operation we get

R3- > R3-2*R2

$$\Rightarrow \begin{bmatrix} 1 & 0 & 0 \\ 0 & 1 & 0 \\ 0 & -2 & 1 \end{bmatrix} \begin{bmatrix} 1 & 0 & 0 \\ -2 & 1 & 0 \\ -3 & 0 & 1 \end{bmatrix} \begin{bmatrix} 1 & 2 & 3 \\ 2 & 3 & 4 \\ 3 & 4 & 6 \end{bmatrix} = \begin{bmatrix} 1 & 2 & 3 \\ 0 & -1 & -2 \\ 0 & 0 & 1 \end{bmatrix}$$

In general, after applying the Row operation the matrix equation will in the form as given below

$$[L_n][L_{n-1}]\ldots[L_3][L_2][L_1][A] = [U]$$

Where

$L_1, L_2, L_3, L_4, \ldots L_n$ are the lower triangular matrices with diagonal elements filled up with ones.
'A' is the actual matrix
'U' is the upper triangular matrix

So the matrix A can be represented as the product of A $= L_1{}^{-1} \ldots L_{n-2}{}^{-1}$ $L_{n-1}{}^{-1} \ldots L_n{}^{-1}$U
In our example

$$L_1 = \begin{bmatrix} 1 & 0 & 0 \\ -2 & 1 & 0 \\ -3 & 0 & 1 \end{bmatrix} \quad L_2 = \begin{bmatrix} 1 & 0 & 0 \\ 0 & 1 & 0 \\ 0 & -2 & 1 \end{bmatrix}$$

$$\Rightarrow L_1{}^{-1} = \begin{bmatrix} 1 & 0 & 0 \\ 2 & 1 & 0 \\ 3 & 0 & 1 \end{bmatrix} \quad L_2{}^{-1} = \begin{bmatrix} 1 & 0 & 0 \\ 0 & 1 & 0 \\ 0 & 2 & 1 \end{bmatrix}$$

Note: In General the inverse of the matrix with the characteristics given below can be obtained by direct observation.
Characteristics:

(a) Diagonal elements filled up ones
(b) One column below diagonal filled up atleast one non-zero elements
(c) All other elements filled up with zeros

Example 1.8. Consider the matrix L_n given below that satisfies all the characteristics mentioned above

$$L_n = \begin{matrix} 1 & 0 & 0 & 0 & 0 \\ a_1 & 1 & 0 & 0 & 0 \\ a_2 & 0 & 1 & 0 & 0 \\ a_3 & 0 & 0 & 1 & 0 \\ a_4 & 0 & 0 & 0 & 1 \end{matrix}$$

The inverse of the matrix L_n is obtained by direct observation as

$$L_n{}^{-1} = \begin{matrix} 1 & 0 & 0 & 0 & 0 \\ -a_1 & 1 & 0 & 0 & 0 \\ -a_2 & 0 & 1 & 0 & 0 \\ -a_3 & 0 & 0 & 1 & 0 \\ -a_4 & 0 & 0 & 0 & 1 \end{matrix}$$

Consider $\begin{bmatrix} 1 & 0 & 0 \\ 0 & 1 & 0 \\ 0 & -2 & 1 \end{bmatrix} \begin{bmatrix} 1 & 0 & 0 \\ -2 & 1 & 0 \\ -3 & 0 & 1 \end{bmatrix} \begin{bmatrix} 1 & 2 & 3 \\ 2 & 3 & 4 \\ 3 & 4 & 6 \end{bmatrix} = \begin{bmatrix} 1 & 2 & 3 \\ 0 & -1 & -2 \\ 0 & 0 & 1 \end{bmatrix}$

$$\Rightarrow \begin{bmatrix} 1 & 2 & 3 \\ 2 & 3 & 4 \\ 3 & 4 & 6 \end{bmatrix} = \begin{bmatrix} 1 & 0 & 0 \\ 2 & 1 & 0 \\ 3 & 0 & 1 \end{bmatrix} \begin{bmatrix} 1 & 0 & 0 \\ 0 & 1 & 0 \\ 0 & 2 & 1 \end{bmatrix} \begin{bmatrix} 1 & 2 & 3 \\ 0 & -1 & -2 \\ 0 & 0 & 1 \end{bmatrix}$$

$$\Rightarrow \begin{bmatrix} 1 & 2 & 3 \\ 2 & 3 & 4 \\ 3 & 4 & 5 \end{bmatrix} = \begin{bmatrix} 1 & 0 & 0 \\ 2 & 1 & 0 \\ 3 & 2 & 1 \end{bmatrix} \begin{bmatrix} 1 & 2 & 3 \\ 0 & -1 & -2 \\ 0 & 0 & 1 \end{bmatrix}$$

The matrix $\begin{bmatrix} 1 & 2 & 3 \\ 0 & -1 & -2 \\ 0 & 0 & 1 \end{bmatrix}$ can be represented as the product of the diagonal matrix and the upper triangular matrix with all the elements in the diagonal are one as given below.

$$\begin{bmatrix} 1 & 2 & 3 \\ 0 & -1 & -2 \\ 0 & 0 & 1 \end{bmatrix} = \begin{bmatrix} 1 & 0 & 0 \\ 0 & -1 & 0 \\ 0 & 0 & 1 \end{bmatrix} \begin{bmatrix} 1 & 2 & 3 \\ 0 & 1 & -2 \\ 0 & 0 & 1 \end{bmatrix}$$

Note: In General the Upper triangular matrix with non-unity diagonal elements can be represented as the product of the diagonal matrix and the Upper triangular matrix with ones in the diagonal as mentioned below

Example 1.9. Consider the Upper triangular matrix with non-unity diagonal elements $\begin{bmatrix} a & b & c \\ 0 & d & e \\ 0 & 0 & f \end{bmatrix}$ which can be represented as the product of

$$\begin{bmatrix} a & b & c \\ 0 & d & e \\ 0 & 0 & f \end{bmatrix} = \begin{bmatrix} a & 0 & 0 \\ 0 & d & 0 \\ 0 & 0 & f \end{bmatrix} \begin{bmatrix} \frac{a}{a} & \frac{b}{a} & \frac{c}{a} \\ 0 & \frac{d}{d} & \frac{e}{d} \\ 0 & 0 & \frac{f}{f} \end{bmatrix}$$

$$= \begin{bmatrix} a & b & c \\ 0 & d & e \\ 0 & 0 & f \end{bmatrix} = \begin{bmatrix} a & 0 & 0 \\ 0 & d & 0 \\ 0 & 0 & f \end{bmatrix} \begin{bmatrix} 1 & \frac{b}{a} & \frac{c}{a} \\ 0 & 1 & \frac{e}{d} \\ 0 & 0 & 1 \end{bmatrix}$$

Thus the invertible matrix $A = \begin{bmatrix} 1 & 2 & 3 \\ 2 & 3 & 4 \\ 3 & 4 & 6 \end{bmatrix}$ is represented as the product of Lower triangular matrix with ones in the diagonal, diagonal matrix and the upper triangular matrix with ones in the diagonal as shown below

$$\begin{bmatrix} 1 & 2 & 3 \\ 2 & 3 & 4 \\ 3 & 4 & 6 \end{bmatrix} = \begin{bmatrix} 1 & 0 & 0 \\ 2 & 1 & 0 \\ 3 & 2 & 1 \end{bmatrix} \begin{bmatrix} 1 & 0 & 0 \\ 0 & -1 & 0 \\ 0 & 0 & 1 \end{bmatrix} \begin{bmatrix} 1 & 2 & 3 \\ 0 & 1 & -2 \\ 0 & 0 & 1 \end{bmatrix}$$

1.4 PLDU Decomposition of an Arbitrary Matrix

In general an arbitrary matrix A can be represented as the product of the permutation matrix (P), Lower triangular matrix with ones in the diagonal, diagonal matrix (D) with non-zero diagonal elements and the Upper triangular matrix with all ones in the diagonal. If the permutation matrix is the identity matrix, then the matrix A is represented as the product of L, D, U (see Section 1.3)

Example 1.10. PLDU Decomposition of the matrix $A = \begin{bmatrix} 1 & 2 & 3 \\ 2 & 4 & 4 \\ 3 & 4 & 6 \end{bmatrix}$

$$\Rightarrow \begin{bmatrix} 1 & 0 & 0 \\ 0 & 1 & 0 \\ 0 & 0 & 1 \end{bmatrix} \begin{bmatrix} 1 & 2 & 3 \\ 2 & 4 & 4 \\ 3 & 4 & 6 \end{bmatrix} = \begin{bmatrix} 1 & 2 & 3 \\ 2 & 4 & 4 \\ 3 & 4 & 6 \end{bmatrix}$$

Note: Row operation on the Identity matrix in the LHS and the same operation done on the RHS will not affect the equality.

R2- > R2-2*R1

R3- > R3-3*R1

$$\Rightarrow \begin{bmatrix} 1 & 0 & 0 \\ -2 & 1 & 0 \\ -3 & 0 & 1 \end{bmatrix} \begin{bmatrix} 1 & 2 & 3 \\ 2 & 4 & 4 \\ 3 & 4 & 6 \end{bmatrix} = \begin{bmatrix} 1 & 2 & 3 \\ 0 & 0 & -2 \\ 0 & -2 & -3 \end{bmatrix}$$

Note that the matrix $\begin{bmatrix} 1 & 2 & 3 \\ 0 & 0 & -2 \\ 0 & -2 & -3 \end{bmatrix}$ cannot be further subjected to mere row operation to obtain the upper triangular format. Hence the following technique using permutation matrix is used.

$$\begin{bmatrix} 1 & 0 & 0 \\ 0 & 0 & 1 \\ 0 & 1 & 0 \end{bmatrix} \begin{bmatrix} 1 & 2 & 3 \\ 2 & 4 & 4 \\ 3 & 4 & 6 \end{bmatrix} = \begin{bmatrix} 1 & 2 & 3 \\ 3 & 4 & 6 \\ 2 & 4 & 4 \end{bmatrix}$$

$$\Rightarrow \begin{bmatrix} 1 & 0 & 0 \\ 0 & 1 & 0 \\ 0 & 0 & 1 \end{bmatrix} \begin{bmatrix} 1 & 0 & 0 \\ 0 & 0 & 1 \\ 0 & 1 & 0 \end{bmatrix} \begin{bmatrix} 1 & 2 & 3 \\ 2 & 4 & 4 \\ 3 & 4 & 6 \end{bmatrix} = \begin{bmatrix} 1 & 2 & 3 \\ 3 & 4 & 6 \\ 2 & 4 & 4 \end{bmatrix}$$

Now applying the Row operation on the identity matrix, we get

R2- > R2-3*R1

R3- > R3-2*R1

$$\begin{bmatrix} 1 & 0 & 0 \\ -3 & 1 & 0 \\ -2 & 0 & 1 \end{bmatrix} \begin{bmatrix} 1 & 0 & 0 \\ 0 & 0 & 1 \\ 0 & 1 & 0 \end{bmatrix} \begin{bmatrix} 1 & 2 & 3 \\ 2 & 4 & 4 \\ 3 & 4 & 6 \end{bmatrix} = \begin{bmatrix} 1 & 2 & 3 \\ 0 & -2 & -3 \\ 0 & 0 & -2 \end{bmatrix}$$

$$\Rightarrow \begin{bmatrix} 1 & 0 & 0 \\ 0 & 0 & 1 \\ 0 & 1 & 0 \end{bmatrix} \begin{bmatrix} 1 & 2 & 3 \\ 2 & 4 & 4 \\ 3 & 4 & 6 \end{bmatrix} = \begin{bmatrix} 1 & 0 & 0 \\ 3 & 1 & 0 \\ 2 & 0 & 1 \end{bmatrix} \begin{bmatrix} 1 & 2 & 3 \\ 0 & -2 & -3 \\ 0 & 0 & -2 \end{bmatrix}$$

Note that the inverse of the matrix $\begin{bmatrix} 1 & 0 & 0 \\ -3 & 1 & 0 \\ -2 & 0 & 1 \end{bmatrix}$ is $\begin{bmatrix} 1 & 0 & 0 \\ 3 & 1 & 0 \\ 2 & 0 & 1 \end{bmatrix}$

$$\Rightarrow \begin{bmatrix} 1 & 2 & 3 \\ 2 & 4 & 4 \\ 3 & 4 & 6 \end{bmatrix} = \begin{bmatrix} 1 & 0 & 0 \\ 0 & 0 & 1 \\ 0 & 1 & 0 \end{bmatrix} \begin{bmatrix} 1 & 0 & 0 \\ 3 & 1 & 0 \\ 2 & 0 & 1 \end{bmatrix} \begin{bmatrix} 1 & 2 & 3 \\ 0 & -2 & -3 \\ 0 & 0 & -2 \end{bmatrix}$$

Note that the inverse of the matrix $\begin{bmatrix} 1 & 0 & 0 \\ 0 & 0 & 1 \\ 0 & 1 & 0 \end{bmatrix}$ is $\begin{bmatrix} 1 & 0 & 0 \\ 0 & 0 & 1 \\ 0 & 1 & 1 \end{bmatrix}$

$$\Rightarrow \begin{bmatrix} 1 & 2 & 3 \\ 2 & 4 & 4 \\ 3 & 4 & 6 \end{bmatrix} = \begin{bmatrix} 1 & 0 & 0 \\ 0 & 0 & 1 \\ 0 & 1 & 0 \end{bmatrix} \begin{bmatrix} 1 & 0 & 0 \\ 3 & 1 & 0 \\ 2 & 0 & 1 \end{bmatrix} \begin{bmatrix} 1 & 0 & 0 \\ 0 & -2 & 0 \\ 0 & 0 & -2 \end{bmatrix} \begin{bmatrix} 1 & 2 & 3 \\ 0 & 1 & 3/2 \\ 0 & 0 & 1 \end{bmatrix}$$

Thus the matrix $\begin{bmatrix} 1 & 2 & 3 \\ 2 & 4 & 4 \\ 3 & 4 & 6 \end{bmatrix}$ is represented as the product of the permutation matrix $\begin{bmatrix} 1 & 0 & 0 \\ 0 & 0 & 1 \\ 0 & 1 & 0 \end{bmatrix}$, Lower triangular matrix $\begin{bmatrix} 1 & 0 & 0 \\ 3 & 1 & 0 \\ 2 & 0 & 1 \end{bmatrix}$ diagonal matrix $\begin{bmatrix} 1 & 0 & 0 \\ 0 & -2 & 0 \\ 0 & 0 & -2 \end{bmatrix}$ and the Upper triangular matrix $\begin{bmatrix} 1 & 2 & 3 \\ 0 & 1 & 3/2 \\ 0 & 0 & 1 \end{bmatrix}$.

1.5 Vector Space and Its Properties

1. Vector space V over a field F is a set with two operation '+' (addition) and '.' (scalar multiplication) such that the following condition holds

$$x,\ y \in V, \text{then } x + y \in V$$
$$x \in V, \text{c} \in F, \text{then} c.x \in \text{F}$$

2. Properties of the vector space:
   (a) Commutative addition
   For all $x, y \in V,\ x + y = y + x$
   (b) Associatively
   $$(x + y) + z = x + (y + z)$$
   (c) Additive identity
   There exists an element $z \in V$ such that $z + x = x$ for all $x \in V$. z is called zero vector
   (d) Additive inverse
   For each $x \in V$, there exists $y \in V$ such that $x + y = z$
   (e) There exists $1 \in F$, such that $1.x = x$ for all $x \in V$
   (f) For all $a, b \in F$ and $x \in V$ $a.(b.x) = (ab).x$
   (g) For all $a \in F$ and $x, y \in V$ $a.(x + y) = a.x + a.y$
   (h) For all $a, b \in F$ and $x \in V$ $(a + b).x = a.x + b.y$
3. Subspace S of the vector space V is a subset of the V such that

$$x, y \in S, \text{then } x + y \in S$$
$$x \in S, \text{c} \in F, \text{then } c.x \in \text{S}$$

Example 1.11. 1. Set of all real numbers $\mathbb{R}$ over the field $\mathbb{R}$ is the vector space.
2. Set of all the vectors of the form $[x, y]$, *where* $x \in \mathbb{R},\ y \in \mathbb{R}$, over the field $\mathbb{R}$ is the vector space which is represented in short as $\mathbb{R}^2$.
   In general $\mathbb{R}^n$ is the vector space over the field $\mathbb{R}$ which is the set of all the vectors of the form $[x_1\ x_2\ x_3\ x_4\ \ldots\ x_n]$, where $x1, x2, x3, \ldots xn \in \mathbb{R}$.

1.6 Linear Independence, Span, Basis and the Dimension of the Vector Space

1.6.1 Linear Independence

Consider the vector space 'V' over the field F. $v_1,\ v_2,\ v_3,\ v_4 \ldots,\ v_n \in V$ are said to be independent if the linear combinations of the above vectors [(i.e.) $\alpha_1\ v_1 +$

$\alpha_2 v_2 + \alpha_3 v_3 + \alpha_4 v_4 \ldots + \alpha_n v_n$, where $\alpha_1, \alpha_1, \alpha_1, \ldots \alpha_1 \in \mathbb{R}]$ is the zero vector only when $\alpha_1 = \alpha_2 = \alpha_3 = \alpha_4 = \cdots \alpha_n = 0$.

Suppose there exists at least one non-zero scalar $\alpha_1, \alpha_2, \alpha_3, \ldots \alpha_n \in \mathbb{R}$ such that $\alpha_1 v_1 + \alpha_2 v_2 + \alpha_3 v_3 + \alpha_4 v_4 \ldots + \alpha_n v_n = \underline{0}$, then any one arbitrary vector among the list $v_1, v_2, v_3, v_4 \ldots, v_n$ can be represented as the linear combinations of the remaining vectors.

For instance if all the scalars $(\alpha_1, \alpha_2, \alpha_3, \ldots \alpha_n \in \mathbb{R})$ are non-zero, then the vector v_1 is represented as the linear combinations of other vectors as shown below.

$$v_1 = \left(-\frac{\alpha_2}{\alpha_1}\right) v_2 + \left(-\frac{\alpha_3}{\alpha_1}\right) v_3 + \left(-\frac{\alpha_4}{\alpha_1}\right) v_4 + \cdots \left(-\frac{\alpha_n}{\alpha_1}\right) v_n$$

This implies that the vector v_1 depends upon the other vectors $v_2,\ v_3,\ v_4 \ldots, v_n$. Similarly any vector in the lists can be represented as the linear combinations of other vectors. This implies that the vectors $v_1,\ v_2,\ v_3,\ v_4 \ldots, v_n$ are dependent vectors.

1.6.2 Span

Consider the vector space 'V' over the field F. Let $v_1,\ v_2,\ v_3,\ v_4 \ldots, v_n \in V_0$ vectors. If all the vectors in the vector space V can be represented as the linear combinations of the above listed vectors, then the listed vectors are called the spanning set of the vector space 'V'.

1.6.3 Basis

Spanning set of the vector space 'V' which consists of the minimum number of independent vectors are called the basis of the vector space 'V'.

1.6.4 Dimension

Number of vectors in the Basis is called the dimension of the vector space 'V'.

1.7 Four Fundamental Vector Spaces of the Matrix

Columns of the matrix can be viewed as the set of column vectors. Similarly Rows of the matrix can be viewed as the set of Row vectors.

1.7.1 Column Space

Column space of the matrix A of size m $\times$ n is the vector space over the field '$\mathbb{R}$' which is the subspace of the vector space $\mathbb{R}^m$. Any vector in the column space of the matrix A can be obtained as the linear combinations of the columns of the matrix. Columns of the matrix A forms the spanning set of the Column space.

1.7.2 Null Space

Null space of the matrix A of size m $\times$ n, represented as N (A), is the vector space over the field '$\mathbb{R}$' which is the subspace of the vector space $\mathbb{R}^n$ such that the $A\ v = \underline{0}$ *for all* $v \in N(A)$.

1.7.3 Row Space

Row space of the matrix A of size m $\times$ n, represented as C(A^T), is the vector space over the field '$\mathbb{R}$' which is the subspace of the vector space $\mathbb{R}^n$, which is basically the Column space of the matrix A^T. Therefore any vector in the Row space of the matrix A can be obtained as linear combinations of the row vectors. Rows of the matrix A forms the spanning set of the Row space.

1.7.4 Left Null Space

Left Null space of the matrix A of size m $\times$ n, represented as N(A^T), is the vector space over the field '$\mathbb{R}$' which is the subspace of the vector space $\mathbb{R}^m$ such that the $A^T\ v = \underline{0}$ *for all* $v \in N(A^T)$.

1.8 Basis of the Four Fundamental Vector Spaces of the Matrix

Example 1.12.

$$A = \begin{bmatrix} 1 & 2 & 3 & 4 \\ 5 & 5 & 7 & 12 \\ 6 & 7 & 10 & 16 \end{bmatrix}$$

1.8.1 Column Space

To compute the column space, we need to find the maximum number of independent columns of the matrix A which is the minimum spanning set (i.e.) Basis of the column space.

Trick to find out maximum number of Independent columns of the matrix A.

Applying the following Row Operation

R2- > R2-5*R1

R3- > R3-6*R1

we get

$$\Rightarrow \begin{bmatrix} 1 & 2 & 3 & 4 \\ 0 & -5 & -8 & -8 \\ 0 & -5 & -8 & -8 \end{bmatrix}$$

R3- > R3-R2

$$\Rightarrow= \begin{bmatrix} 1 & 2 & 3 & 4 \\ 0 & -5 & -8 & -8 \\ 0 & 0 & 0 & 0 \end{bmatrix}$$

$$\Rightarrow \begin{bmatrix} \mathbf{1} & 2 & 3 & 4 \\ 0 & \mathbf{-5} & -8 & -8 \\ 0 & 0 & 0 & 0 \end{bmatrix}$$

The above is the Row Reduced Echelon Form (RREF). The Bold numbers in the above matrix are called pivot elements and the corresponding columns are called pivot columns.

$$\begin{bmatrix} \mathbf{1} & 2 & 3 & 4 \\ 0 & -5 & \mathbf{-8} & -8 \\ 0 & 0 & 0 & 0 \end{bmatrix}$$

The bold numbers mentioned in the above matrix can also be treated as the pivot numbers and the corresponding columns are called pivot columns.

$$A = \begin{bmatrix} \mathbf{1} & 2 & 3 & 4 \\ 0 & -5 & -8 & \mathbf{-8} \\ 0 & 0 & 0 & 0 \end{bmatrix}$$

Similarly the bold numbers mentioned in the above matrix can be treated as the pivot numbers and the corresponding columns are called pivot columns.

Pivot columns are independent to each other and it is the maximum number of independent columns of the RREF matrix. It can be shown that the corresponding columns of the original matrix A are independent vectors.

Check for independent vectors

The Linear combinations of the independent vectors is equal to zero vector only when the scaling factors are identically zeros.

(i.e.)

$$\alpha_1 \begin{bmatrix} 1 \\ 5 \\ 6 \end{bmatrix} + \alpha_2 \begin{bmatrix} 2 \\ 5 \\ 7 \end{bmatrix} = \begin{bmatrix} 0 \\ 0 \\ 0 \end{bmatrix} \Rightarrow \alpha_1 = \alpha_2 = 0$$

Rewriting the above equations in the standard Linear equation form is as shown below.

$$\alpha_1 + 2\alpha_2 = 0\text{-----------------------(1)}$$
$$5\alpha_1 + 5\alpha_2 = 0\text{---------------------(2)}$$
$$6\alpha_1 + 7\alpha_2 = 0\text{---------------------(3)}$$

(2)–5*(1) and (3)–6*(1) gives the following equations

$$\alpha_1 + 2\alpha_2 = 0\text{-------------------------(1)}$$
$$-5\alpha_2 = 0\text{-----------------------(2)}$$
$$-5\alpha_2 = 0\text{-----------------------(3)}$$
$$\Rightarrow \alpha_1 = \alpha_2 = 0$$

The operation performed above is equivalent to the Row operation of the original matrix A. **Thus columns of the original matrix corresponding to the pivot columns are independent to each other**.

Hence the corresponding columns in the original matrix form the column space of the matrix A. Thus column space of the matrix A is represented as set of vectors

$$\alpha_1 \begin{bmatrix} 1 \\ 5 \\ 6 \end{bmatrix} + \alpha_2 \begin{bmatrix} 2 \\ 5 \\ 7 \end{bmatrix}, \text{ where } \alpha_1, \alpha_1 \in \mathbb{R}$$

Also note that the dimension of the column space is 2.

Null space of the matrix A is obtained as follows.

$$A = \begin{bmatrix} 1 & 2 & 3 & 4 \\ 5 & 5 & 7 & 12 \\ 6 & 7 & 10 & 16 \end{bmatrix}$$

Row and the column operation of the matrix can be viewed as the matrix multiplication of the particular matrix with the matrix itself as described below.

$$\Rightarrow \begin{bmatrix} 1 & 0 & 0 \\ 0 & 1 & 0 \\ 0 & 0 & 1 \end{bmatrix} \begin{bmatrix} 1 & 2 & 3 & 4 \\ 5 & 5 & 7 & 12 \\ 6 & 7 & 10 & 16 \end{bmatrix} = \begin{bmatrix} 1 & 2 & 3 & 4 \\ 5 & 5 & 7 & 12 \\ 6 & 7 & 10 & 16 \end{bmatrix}$$

Applying the Row operation on the Identity matrix in the LHS and the matrix in the RHS, we get

R2- > R2-5*R1

R3- > R3-6*R1

$$\begin{bmatrix} 1 & 0 & 0 \\ -5 & 1 & 0 \\ -6 & 0 & 1 \end{bmatrix} \begin{bmatrix} 1 & 2 & 3 & 4 \\ 5 & 5 & 7 & 12 \\ 6 & 7 & 10 & 16 \end{bmatrix} = \begin{bmatrix} 1 & 2 & 3 & 4 \\ 0 & -5 & -8 & -8 \\ 0 & -5 & -8 & -8 \end{bmatrix}$$

$$\Rightarrow \begin{bmatrix} 1 & 0 & 0 \\ 0 & 1 & 0 \\ 0 & 0 & 1 \end{bmatrix} \begin{bmatrix} 1 & 0 & 0 \\ -5 & 1 & 0 \\ -6 & 0 & 1 \end{bmatrix} \begin{bmatrix} 1 & 2 & 3 & 4 \\ 5 & 5 & 7 & 12 \\ 6 & 7 & 10 & 16 \end{bmatrix} = \begin{bmatrix} 1 & 2 & 3 & 4 \\ 0 & -5 & -8 & -8 \\ 0 & -5 & -8 & -8 \end{bmatrix}$$

Again applying the Row operation on the Identity matrix in the LHS and the matrix in the RHS, we get

$$\begin{bmatrix} 1 & 0 & 0 \\ 0 & 1 & 0 \\ 0 & -1 & 1 \end{bmatrix} \begin{bmatrix} 1 & 0 & 0 \\ -5 & 1 & 0 \\ -6 & 0 & 1 \end{bmatrix} \begin{bmatrix} 1 & 2 & 3 & 4 \\ 5 & 5 & 7 & 12 \\ 6 & 7 & 10 & 16 \end{bmatrix} = \begin{bmatrix} 1 & 2 & 3 & 4 \\ 0 & -5 & -8 & -8 \\ 0 & 0 & 0 & 0 \end{bmatrix}$$

Thus an arbitrary vector $\begin{bmatrix} x1 \\ x2 \\ x3 \\ x4 \end{bmatrix}$ if multiplied with the matrix $\begin{bmatrix} 1 & 2 & 3 & 4 \\ 5 & 5 & 7 & 12 \\ 6 & 7 & 10 & 16 \end{bmatrix}$ gives the zero vector, then the same vector $\begin{bmatrix} x1 \\ x2 \\ x3 \\ x4 \end{bmatrix}$ when multiplied with the matrix $\begin{bmatrix} 1 & 2 & 3 & 4 \\ 0 & -5 & -8 & -8 \\ 0 & 0 & 0 & 0 \end{bmatrix}$ gives the zero vector. In other words the Null space of the matrix $\begin{bmatrix} 1 & 2 & 3 & 4 \\ 5 & 5 & 7 & 12 \\ 6 & 7 & 10 & 16 \end{bmatrix}$ is same as that of the Null space of the matrix after performing Row operation. (i.e.)

Null space of the matrix $\begin{bmatrix} 1 & 2 & 3 & 4 \\ 0 & -5 & -8 & -8 \\ 0 & 0 & 0 & 0 \end{bmatrix}$.

Null space of the matrix $\begin{bmatrix} 1 & 2 & 3 & 4 \\ 0 & -5 & -8 & -8 \\ 0 & 0 & 0 & 0 \end{bmatrix}$ is determined as follows.

$$\begin{bmatrix} \mathbf{1} & 2 & 3 & 4 \\ 0 & \mathbf{-5} & -8 & -8 \\ 0 & 0 & 0 & 0 \end{bmatrix} \begin{bmatrix} x1 \\ x2 \\ x3 \\ x4 \end{bmatrix} = \begin{bmatrix} 0 \\ 0 \\ 0 \end{bmatrix}$$

Bold numbers of the above matrix is treated as pivot columns. Remaining columns are called free variable columns. The corresponding variables $x1$ and $x2$ are called pivot variables and the variables $x3$ and $x4$ are called free variables.

Representing the pivot variables in terms of free variables, we get

$$x1 = -2 * \left[-\frac{8}{5} * x3 - \frac{8}{5} * x4 \right] - 3 * x3 - 4 * x4$$

$$x1 = \frac{16}{5} * x3 + \frac{16}{5} * x4 - 3 * x3 - 4 * x4$$

$$x1 = \frac{1}{5} * x3 - \frac{4}{5} * x4$$

$$x2 = -\frac{8}{5} * x3 - \frac{8}{5} * x4$$

$$x3 = x3$$

$$x4 = x4$$

Thus the set of vectors as shown below form the Null space of the matrix A.

$$\begin{bmatrix} x1 \\ x2 \\ x3 \\ x4 \end{bmatrix} = x3 \begin{bmatrix} \frac{1}{5} \\ -\frac{8}{5} \\ 1 \\ 0 \end{bmatrix} + x4 \begin{bmatrix} -\frac{4}{5} \\ -\frac{8}{5} \\ 0 \\ 1 \end{bmatrix}, x3, x4 \in \mathbb{R}$$

As shown above, the number of independent columns to represent the Null space is two. Hence dimension of the Null space of the matrix A is given as two.

Row space and the Left Null space of the matrix A is obtained as the Column space and the Null space of the matrix A^T as shown below.

$$\mathrm{A} = \begin{bmatrix} 1 & 2 & 3 & 4 \\ 5 & 5 & 7 & 12 \\ 6 & 7 & 10 & 16 \end{bmatrix} \quad A^T = \begin{bmatrix} 1 & 5 & 6 \\ 2 & 5 & 7 \\ 3 & 7 & 10 \\ 4 & 12 & 16 \end{bmatrix}$$

Applying Row operation,

R2- > R2-2* R1

R3- > R3-3* R1

R4- > R4-4* R1

We get,

$$\begin{bmatrix} 1 & 5 & 6 \\ 0 & -5 & -5 \\ 0 & -8 & -8 \\ 0 & -8 & -8 \end{bmatrix}$$

Further applying the Row operation

R3- > R3-(5/8)* R2

R4- > R4-(5/8)* R2

We get,

$$\begin{bmatrix} \mathbf{1} & 5 & 6 \\ 0 & \mathbf{-5} & -5 \\ 0 & 0 & 0 \\ 0 & 0 & 0 \end{bmatrix}$$

The Bold numbers in the above matrix are called pivots and the corresponding columns are called pivot columns. The columns of the original matrix corresponding to the pivot columns are the maximum number of independent columns of the matrix A. Hence they are the basis of the column space of the matrix A^T or the Row space of the matrix A.

Thus the Row space of the matrix A is represented as the set as given below.

$$\alpha_1 \begin{bmatrix} 1 \\ 2 \\ 3 \\ 4 \end{bmatrix} + \alpha_2 \begin{bmatrix} 5 \\ 5 \\ 7 \\ 12 \end{bmatrix}, \alpha_1, \alpha_2 \in \mathbb{R}$$

The dimension of the Row space of the matrix A is 2.

Similarly the Left Null space of the matrix A is obtained as the Null space of the matrix A^T

The matrix A^T after subjected to Row operation is as shown below.

$$\begin{bmatrix} \mathbf{1} & 5 & 6 \\ 0 & \mathbf{-5} & -5 \\ 0 & 0 & 0 \\ 0 & 0 & 0 \end{bmatrix}$$

Bold letters mentioned above in the matrix are the pivot elements and the corresponding columns are the pivot columns. Other column is called free variable column. Null space of the matrix A^T is same as that of the Null space of the matrix A^T after subjected to Row operation. Thus the vector space of the form $\begin{bmatrix} x1 \\ x2 \\ x3 \end{bmatrix}$ is required such that $\begin{bmatrix} \mathbf{1} & 5 & 6 \\ 0 & \mathbf{-5} & -5 \\ 0 & 0 & 0 \\ 0 & 0 & 0 \end{bmatrix} \begin{bmatrix} x1 \\ x2 \\ x3 \end{bmatrix} = \begin{bmatrix} 0 \\ 0 \\ 0 \\ 0 \end{bmatrix}$ to obtain the Null space of the matrix A^T. The variables $x1$ and $x2$ are called pivot variables and the variable $x3$ is called free variable.

Representing the pivot variables in terms of free variables, we get the following.

$$
\begin{aligned}
x1 + 5 * x2 \; + \; 6 * x3 &= 0\text{- -(1)} \\
-5 * x2 \; - \; 5 * x3 &= 0\text{- - - - - - - - - - - - - - - - - - - -(2)} \\
\text{Equation (2)} \Rightarrow x2 &= -x3 \\
\text{Equation (1)} \Rightarrow x1 &= -5 * x2 - 6 * x3 \\
\Rightarrow x1 &= 5 * x3 - 6 * x3 = -x3
\end{aligned}
$$

Thus the Basis of the Left Null space of the matrix A, which is the Basis of the Null space of the matrix A^T is given below.

$$
\begin{bmatrix} -x3 \\ -x3 \\ x3 \end{bmatrix} \forall x3 \in \mathbb{R}
$$

$$
= \begin{bmatrix} -1 \\ -1 \\ 1 \end{bmatrix} \alpha \; \forall \, \alpha \; \in \mathbb{R}
$$

Note that the dimension of the Left Null space of the matrix A is one.

1.9 Observations on Results of the Example 1.12

In the Example 1.12, Column space, Null space, Left Column space (Row space) and the Left Null space of the matrix A $= \begin{bmatrix} 1 & 2 & 3 & 4 \\ 5 & 5 & 7 & 12 \\ 6 & 7 & 10 & 16 \end{bmatrix}$ is obtained as the following

1.9.1 Column Space

$$\alpha_1 \begin{bmatrix} 1 \\ 5 \\ 6 \end{bmatrix} + \alpha_2 \begin{bmatrix} 2 \\ 5 \\ 7 \end{bmatrix}, \text{where } \alpha_1, \alpha_1 \in \mathbb{R}$$

The dimension of the column space is two. Also note that the Column space of the matrix A sized 3×4 is the subspace of the vector space $\mathbb{R}^3$.

1.9.2 Null Space

$$\begin{bmatrix} x1 \\ x2 \\ x3 \\ x4 \end{bmatrix} = x3 \begin{bmatrix} \frac{1}{5} \\ -\frac{8}{5} \\ 1 \\ 0 \end{bmatrix} + x4 \begin{bmatrix} -\frac{4}{5} \\ -\frac{8}{5} \\ 0 \\ 1 \end{bmatrix} \quad \forall x3, x4 \in \mathbb{R}$$

The dimension of the Null space is two. Also note that the Null pace of the matrix A sized 3×4 is subspace of the vector space $\mathbb{R}^4$.

1.9.3 Left Column Space (Row Space)

$$\alpha_1 \begin{bmatrix} 1 \\ 2 \\ 3 \\ 4 \end{bmatrix} + \alpha_2 \begin{bmatrix} 5 \\ 5 \\ 7 \\ 12 \end{bmatrix} \quad \forall\, \alpha_1, \alpha_2 \in \mathbb{R}$$

The dimension of the Row space is two. Also note that the Row space of the matrix A sized 3×4 is subspace of the vector space $\mathbb{R}^4$.

1.9.4 Left Null Space

$$\begin{bmatrix} -1 \\ -1 \\ 1 \end{bmatrix} \alpha \;\forall\, \alpha \in \mathbb{R}$$

The dimension of the Left Null space is one. Also note that the Left Null space of the matrix A sized 3×4 is subspace of the vector space $\mathbb{R}^3$.

1.9.5 Observation

1. The column space and the Left Null space of the matrix A are the subspaces of the $\mathbb{R}^3$, where 3 is the number of rows of the matrix A.
2. The Null space and the Left column space of the matrix A are the subspaces of the $\mathbb{R}^4$, where 4 is the number of columns of the matrix A.
3. Dimension of the column space of the matrix A + Dimension of the Null space of the matrix A $= 2 + 2 = 4 =$ Number of Columns of the matrix A.
4. Dimension of the Left Null space of the matrix A+ Dimension of the Left Column space (Row space) of the matrix A $= 1 + 2 = 4 =$ Number of Rows of the matrix A.
5. The Column space of the matrix A and the Left Null space of the matrix A are orthogonal to each other (i.e.) any vector a $\in$ C(A) and b $\in N(A^T)$ satisfies the condition a^Tb $= b^T$a $= \underline{0}$.

   Proof. Let $y \in \mathrm{N}(A^T) \Rightarrow A^T y = \underline{0}$. Taking transpose on both sides, we get $y^T A = \underline{0} \Rightarrow y^T Ax = \underline{0}$ (Multiplying arbitrary vector on both sides).
   Note that $Ax \in C(A)$. Hence proved.

6. The Row of the matrix A and the Null space of the matrix A are orthogonal to each other.

   Proof. Let $y \in$ N(A) $\Rightarrow$ A$y = \underline{0}$. Taking transpose on both sides, we get $y^T A^T = \underline{0} \Rightarrow y^T A^T x = \underline{0}$ (Multiplying arbitrary vector x on both sides). Note that $A^T x \in C(A^T)$. Hence proved.

In general for the matrix A of size m $\times$ n dim($C(A)$) + dim ($N(A)$) = n. This is known as Rank-Nullity Theorem.

Dimension of the Column space of the matrix A is the maximum number of independent columns (pivot columns) of the matrix A. Let it be 'r'. From the procedure of the determining the Null space of the matrix A (Representing all the variables in terms of free variables), it can be shown that dimension of the Null space of the matrix A is equal to the number of free variable columns of the matrix A. This is equal to n-r, where 'n' is the total number of columns of the matrix A and the 'r' is the number of pivot columns. Thus dim(C(A)) + dim(R(A)) = n.

1.10 Vector Representation with Different Basis

Example 1.13. $\begin{bmatrix}3\\4\end{bmatrix}_{\begin{bmatrix}1\\0\end{bmatrix}\begin{bmatrix}0\\1\end{bmatrix}} \in \mathbb{R}^2$ is the vector represented with respect to the basis $\begin{bmatrix}1\\0\end{bmatrix}\begin{bmatrix}0\\1\end{bmatrix}$. (i.e.) The vector $\begin{bmatrix}3\\4\end{bmatrix}_{\begin{bmatrix}1\\0\end{bmatrix}\begin{bmatrix}0\\1\end{bmatrix}} = 3^* \begin{bmatrix}1\\0\end{bmatrix}_{\begin{bmatrix}1\\0\end{bmatrix}\begin{bmatrix}0\\1\end{bmatrix}} + 4^* \begin{bmatrix}0\\1\end{bmatrix}_{\begin{bmatrix}1\\0\end{bmatrix}\begin{bmatrix}0\\1\end{bmatrix}}$

Similarly $\begin{bmatrix}5\\6\end{bmatrix}_{\begin{bmatrix}1\\1\end{bmatrix}\begin{bmatrix}1\\-1\end{bmatrix}} \in \mathbb{R}^2$ is the vector represented with respect to the basis $\begin{bmatrix}1\\1\end{bmatrix}\begin{bmatrix}1\\-1\end{bmatrix}$. (i.e.) The vector $\begin{bmatrix}5\\6\end{bmatrix}_{\begin{bmatrix}1\\1\end{bmatrix}\begin{bmatrix}1\\-1\end{bmatrix}} = 5^* \begin{bmatrix}1\\1\end{bmatrix}_{\begin{bmatrix}1\\1\end{bmatrix}\begin{bmatrix}1\\-1\end{bmatrix}} + 6^* \begin{bmatrix}1\\-1\end{bmatrix}_{\begin{bmatrix}1\\1\end{bmatrix}\begin{bmatrix}1\\-1\end{bmatrix}}$

The vector $\begin{bmatrix}5\\6\end{bmatrix}_{\begin{bmatrix}1\\1\end{bmatrix}\begin{bmatrix}1\\-1\end{bmatrix}} = 5^* \begin{bmatrix}1\\1\end{bmatrix}_{\begin{bmatrix}1\\1\end{bmatrix}\begin{bmatrix}1\\-1\end{bmatrix}} + 6^* \begin{bmatrix}1\\-1\end{bmatrix}_{\begin{bmatrix}1\\1\end{bmatrix}\begin{bmatrix}1\\-1\end{bmatrix}}$

Also the vector $\begin{bmatrix}1\\1\end{bmatrix}_{\begin{bmatrix}1\\1\end{bmatrix}\begin{bmatrix}1\\-1\end{bmatrix}}$ is represented as the linear combinations of $\begin{bmatrix}1\\0\end{bmatrix}$ and $\begin{bmatrix}0\\1\end{bmatrix}$ as $1 * \begin{bmatrix}1\\0\end{bmatrix} + 1 * \begin{bmatrix}0\\1\end{bmatrix}$

Similarly the vector $\begin{bmatrix}1\\-1\end{bmatrix}_{\begin{bmatrix}1\\1\end{bmatrix}\begin{bmatrix}1\\-1\end{bmatrix}}$ is represented as the linear combinations of $\begin{bmatrix}1\\0\end{bmatrix}$ and $\begin{bmatrix}0\\1\end{bmatrix}$ as $1 * \begin{bmatrix}1\\0\end{bmatrix} + (-1) * \begin{bmatrix}0\\1\end{bmatrix}$

Thus the vector $\begin{bmatrix}5\\6\end{bmatrix}_{\begin{bmatrix}1\\1\end{bmatrix}\begin{bmatrix}1\\-1\end{bmatrix}}$ is represented as the following

$$5^* \left(1 * \begin{bmatrix}1\\0\end{bmatrix} + 1 * \begin{bmatrix}0\\1\end{bmatrix}\right) + 6^* \left(1 * \begin{bmatrix}1\\0\end{bmatrix} + (-1) * \begin{bmatrix}0\\1\end{bmatrix}\right)$$

$$= (5 * 1 + 6 * 1) * \begin{bmatrix}1\\0\end{bmatrix} + (5 * 1 + 6 * 1) * \begin{bmatrix}0\\1\end{bmatrix}$$

$$= \left\{\begin{bmatrix}1 & 1\\1 & -1\end{bmatrix}\begin{bmatrix}5\\6\end{bmatrix}\right\}_{\begin{bmatrix}1\\0\end{bmatrix},\begin{bmatrix}0\\1\end{bmatrix}}$$

$$\begin{bmatrix}5\\6\end{bmatrix}_{\begin{bmatrix}1\\1\end{bmatrix}\begin{bmatrix}1 & -1\\1 & -1\end{bmatrix}} = \left\{\begin{bmatrix}1 & 1\\1 & -1\end{bmatrix}\begin{bmatrix}5\\6\end{bmatrix}\right\}_{\begin{bmatrix}1\\0\end{bmatrix},\begin{bmatrix}0\\1\end{bmatrix}}$$

$$\Rightarrow \begin{bmatrix}5\\6\end{bmatrix}_{\begin{bmatrix}1\\1\end{bmatrix}\begin{bmatrix}1\\-1\end{bmatrix}} = \begin{bmatrix}11\\-1\end{bmatrix}_{\begin{bmatrix}1\\0\end{bmatrix},\begin{bmatrix}0\\1\end{bmatrix}}$$

Similarly $\begin{bmatrix}5\\6\end{bmatrix}_{\begin{bmatrix}1\\0\end{bmatrix}\begin{bmatrix}0\\1\end{bmatrix}} = \left\{\begin{bmatrix}1 & 1\\1 & -1\end{bmatrix}^{-1}\begin{bmatrix}5\\6\end{bmatrix}\right\}_{\begin{bmatrix}1\\1\end{bmatrix},\begin{bmatrix}1\\-1\end{bmatrix}}$

1.11 Linear Transformation of the Vector

T: V->U is the Linear transformation such that any vector in the vector space 'V' is mapped to another vector that lies in the vector space 'U'. There exists the transformation matrix to perform this operation. The vector space V can also be equal to the vector space U.

Example 1.14. T: $\mathbb{R}^2$->$\mathbb{R}^2$ such that any vector $\underline{x} \in \mathbb{R}^2$ is mapped to another vector $\underline{y} \in \mathbb{R}^2$ using the relation $\underline{y} = \begin{bmatrix}-1 & 0\\0 & -1\end{bmatrix}\underline{x}$. This is the Linear operation of the image reflection about the origin. Note that the vectors $\underline{x}$ and $\underline{y}$ are represented with respect to the standard basis. Also note that the transformation matrix is with respect to the standard basis.

$\mathbb{R}^2$ Vector points plotted in the 2D plane before and after transformation are given below (Fig. 1.2).

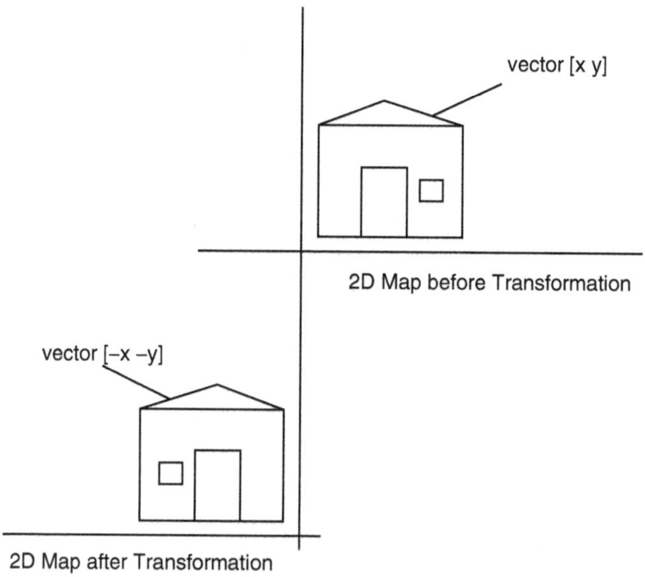

Fig. 1.2 Illustration of the linear transformation of the vector

1.11.1 Trick to Compute the Transformation Matrix

Identify and note down the transformed vectors for the standard basis $\begin{bmatrix}1\\0\end{bmatrix}$ and $\begin{bmatrix}1\\0\end{bmatrix}$.

For the example mentioned above $\begin{bmatrix}-1\\0\end{bmatrix}$ and $\begin{bmatrix}0\\-1\end{bmatrix}$ are the transformed vectors corresponding to the standard basis $\begin{bmatrix}1\\0\end{bmatrix}$ and $\begin{bmatrix}1\\0\end{bmatrix}$ respectively. The transformation matrix corresponding to the above transformation is obtained by representing the transformed vectors column wise. For the above mentioned example, the transformation matrix is given as

$$\begin{bmatrix}-1 & 0\\0 & -1\end{bmatrix}$$

This is same as the one given in the Example 1.14.

1.12 Transformation Matrix with Different Basis

Consider the transformation matrix with respect to the standard basis as described below

$$\begin{bmatrix}-1 & 0\\0 & -1\end{bmatrix}$$

Consider the vector $\begin{bmatrix}5\\6\end{bmatrix}_{\begin{bmatrix}1\\1\end{bmatrix}\begin{bmatrix}1\\-1\end{bmatrix}} = \left\{\begin{bmatrix}1 & 1\\1 & -1\end{bmatrix}\begin{bmatrix}5\\6\end{bmatrix}\right\}_{\begin{bmatrix}1\\0\end{bmatrix},\begin{bmatrix}0\\1\end{bmatrix}}$ represented with respect to the basis $\begin{bmatrix}1\\1\end{bmatrix}\begin{bmatrix}1\\-1\end{bmatrix}$ and $\begin{bmatrix}1\\0\end{bmatrix},\begin{bmatrix}0\\1\end{bmatrix}$ respectively.

The vector in the standard basis (i.e.) $\left\{\begin{bmatrix}1 & 1\\1 & -1\end{bmatrix}\begin{bmatrix}5\\6\end{bmatrix}\right\}_{\begin{bmatrix}1\\0\end{bmatrix},\begin{bmatrix}0\\1\end{bmatrix}}$ is transformed into another vector in the standard basis using the transformation matrix $\begin{bmatrix}-1 & 0\\0 & -1\end{bmatrix}$ as

$$\left\{\begin{bmatrix}-1 & 0\\0 & -1\end{bmatrix}\begin{bmatrix}1 & 1\\1 & -1\end{bmatrix}\begin{bmatrix}5\\6\end{bmatrix}\right\}_{\begin{bmatrix}1\\0\end{bmatrix},\begin{bmatrix}0\\1\end{bmatrix}}$$

The above vector in standard basis is equivalent to the vector given below with respect to the basis $\begin{bmatrix}1\\1\end{bmatrix}\begin{bmatrix}1\\-1\end{bmatrix}$

$$\left\{\begin{bmatrix}1 & 1\\ 1 & -1\end{bmatrix}^{-1}\begin{bmatrix}-1 & 0\\ 0 & -1\end{bmatrix}\begin{bmatrix}1 & 1\\ 1 & -1\end{bmatrix}\begin{bmatrix}5\\ 6\end{bmatrix}\right\}_{\begin{bmatrix}1\\ 1\end{bmatrix},\begin{bmatrix}1\\ -1\end{bmatrix}}$$

Note that the vector $\begin{bmatrix}5\\ 6\end{bmatrix}$ and the transformed vector are represented with respect to the basis $\begin{bmatrix}1\\ 1\end{bmatrix},\begin{bmatrix}1\\ -1\end{bmatrix}$

The matrix $\begin{bmatrix}-1 & 0\\ 0 & -1\end{bmatrix}$ is the transformation matrix with respect to the standard basis and the matrix $\left\{\begin{bmatrix}1 & 1\\ 1 & -1\end{bmatrix}^{-1}\begin{bmatrix}-1 & 0\\ 0 & -1\end{bmatrix}\begin{bmatrix}1 & 1\\ 1 & -1\end{bmatrix}\right\}_{\begin{bmatrix}1\\ 1\end{bmatrix},\begin{bmatrix}1\\ -1\end{bmatrix}}$ is the corresponding transformation matrix with respect to the basis $\begin{bmatrix}1\\ 1\end{bmatrix},\begin{bmatrix}1\\ -1\end{bmatrix}$ These matrices are called similar matrices. In general similar matrix of the matrix A is obtained using invertible matrix M as $M^{-1}AM$.

1.13 Orthogonality

1.13.1 Basic Definitions and Results

Inner product: Inner product of the vectors $\underline{x}, \underline{y} \in \mathbb{R}^n$ is defined as $\underline{x}^T \underline{y} =$

$$\begin{bmatrix}x1\\ x2\\ .\\ .\\ .\\ xn\end{bmatrix}\begin{bmatrix}y1y2 & \ldots & yn\end{bmatrix} = x1 * y1 + x2 * y2 + x3 * y3 \ldots + xn * yn$$

Orthogonal vectors: Two vectors $\underline{x}$ and $\underline{y}$ are said to be orthogonal if $\underline{x}^T \underline{y} = \underline{0}$.

Orthogonal basis: Basis B is said to be orthogonal if $\underline{x}^T \underline{y} = \underline{0}\ \forall\ \underline{x}, \underline{y} \in B,\ \underline{x} \neq \underline{y}$

Orthonormal basis: Basis B is said to be orthonormal basis if $\underline{x}^T \underline{y} = \underline{0}$ and $\underline{x}^T \underline{x} = 1\ \forall\ \underline{x}, \underline{y} \in B, \underline{x} \neq \underline{y}$

Example. Standard basis for the vector space $\mathbb{R}^n$

A set of mutually orthogonal non-zero vectors are linearly independent. But set of independent vectors need not be orthogonal vectors.

Orthogonality of subspaces: Consider two subspaces S $\subseteq \mathbb{R}^n$ and T $\subseteq \mathbb{R}^n$. The two subspace S and T are said to be orthogonal if $\underline{x}^T \underline{y} = \underline{0}\ \forall\ \underline{x}, \in S$ and $\underline{y} \in T, \underline{x} \neq \underline{y}$

Example. 1. Column space of the arbitrary matrix A and the Left Null space of the matrix A are the orthogonal spaces.
2. Row space of the arbitrary matrix A and the Null space of the matrix A are orthogonal spaces.

1.13.2 Orthogonal Complement

The Vector space 'V' that consists of the all the vectors that are orthogonal to the vector space 'S' is called the orthogonal complement of the vector space 'S'. It is denoted as V = $S^{\perp}$.

Example 1.15. 1. Column space of the arbitrary matrix A and the Left Null space of the matrix A are the orthogonal complement.
Row space of the arbitrary matrix A and the Null space of the matrix A are the orthogonal complement.

1.14 System of Linear Equation

System of Linear Equation can be viewed as the problem of estimating the value of $c_i{}'S$, (i.e.) $\boldsymbol{c_1}, \boldsymbol{c_2}, \boldsymbol{c_3}, \ldots \boldsymbol{c_m} \in \mathbb{R}$ such that $\boldsymbol{c_1}{}^*\underline{\boldsymbol{X}_1} + \boldsymbol{c_2}{}^*\underline{\boldsymbol{X}_2} + \ldots \boldsymbol{c_m}{}^*\underline{\boldsymbol{X}_m} = \underline{\boldsymbol{b}}$.

$$\text{where } \underline{X_i} = \begin{bmatrix} x_{1i} \\ x_{2i} \\ x_{3i} \\ . \\ . \\ x_{ni} \end{bmatrix} \text{ and } \underline{b} = \begin{bmatrix} b_1 \\ b_2 \\ b_3 \\ . \\ . \\ b_n \end{bmatrix} \text{ and } x_{ij}, b_j \in \mathbb{R}$$

The Expanded representation of the above mentioned equation is represented as follows.

$$c_1 x_{11} + c_2 x_{12} + c_3 x_{13} + \cdots c_m x_{1m} = b1$$
$$c_1 x_{21} + c_2 x_{22} + c_3 x_{23} + \cdots c_m x_{2m} = b2$$
$$c_1 x_{31} + c_2 x_{32} + c_3 x_{33} + \cdots c_m x_{3m} = b3$$
$$\ldots$$

$$\cdots$$
$$\cdots$$
$$c_1x_{n1} + c_2x_{n2} + c_3x_{n3} + \cdots c_mx_{nm} = b_n$$

Thus the above equation can also be represented as

$$\begin{bmatrix} x_{11} & x_{12} & x_{13} & \cdots & x_{1m} \\ x_{21} & x_{22} & x_{23} & \cdots & x_{2m} \\ x_{31} & x_{32} & x_{33} & \cdots & x_{3m} \\ \cdots & \cdots & \cdots & \cdots & \cdots \\ x_{n1} & x_{n2} & x_{n3} & \cdots & x_{nm} \end{bmatrix} \begin{bmatrix} c_1 \\ c_2 \\ c_3 \\ \cdots \\ c_m \end{bmatrix} = \begin{bmatrix} b_1 \\ b_2 \\ b_3 \\ \cdots \\ b_m \end{bmatrix}$$
$$= [\mathrm{X}]\underline{c} = \underline{b}$$

1.15 Solutions for the System of Linear Equation [A] $\underline{x} = \underline{b}$

Note:

Let the size of the matrix 'A' be m × n. $x \in \mathbb{R}^n, b \in \mathbb{R}^m$. Basis of the vector space $\mathbb{R}^n$ can be obtained as the concatenation of the basis vectors in the Row space of the matrix A with the Basis of the Null space of the matrix 'A'.

Let $\{b_1, b_2, \ldots b_r\}$ be the basis of the Row space of the matrix A. Also $\{b_{r+1}, b_{r+2}, \ldots b_n\}$ be the basis of the Null space of the matrix A. Then the basis of the vector space $\mathbb{R}^n$ is given a $\{b_1, b_2, \ldots b_n\}$.

Consider any arbitrary vector 'v' in the vector space $\mathbb{R}^n$, which can be represented as the linear combinations of the above mentioned basis vectors as given below.

$$\begin{aligned} v &= \alpha_1 b_1 + \alpha_2 b_2 + \alpha_3 b_3 + \cdots \alpha_r b_r + \alpha_{r+1} b_{r+1} + \alpha_{r+2} b_{r+2} + \cdots + \alpha_n b_n \\ &= \alpha_1 \mathbf{b_1} + \alpha_2 \mathbf{b_2} + \alpha_3 \mathbf{b_3} + \cdots \alpha_r \mathbf{b}_r) + (\alpha_{r+1} b_{r+1} + \alpha_{r+2} b_{r+2} + \cdots + a_n b_n) \\ &= \boldsymbol{vr} + \boldsymbol{vn} \end{aligned}$$

Thus any vector 'v' in the vector space $\mathbb{R}^n$ can be represented as the direct summation of the vector from the Row space 'vr' and the vector from the Null space 'vn'.

Similarly any vector in the vector space $\mathbb{R}^m$ can be represented as the summation of the vector from the column space and the vector from the Left Null space.

Case 1: If the vector $\underline{b}$ lies in the column space of the matrix A

The solution that exists in the vector space $\mathbb{R}^n$ can be represented as the direct sum of the vector from the row space and the vector from the null space. The vector which is obtained from the row space is unique. But any vector from the null space can be chosen to add with the one chosen from the row space to get the solution.

Thus in general, there exists infinite number of solutions for the system of Linear equation. But, if the Null space of the matrix 'A' is $\underline{0}$, then the unique solution occurs for the equation of Linear Equation **[A]** $\underline{\boldsymbol{x}} = \underline{\boldsymbol{b}}$. In this case the unique solution is the vector obtained from the row space

Example 1.16. Consider the Linear Equation represented in the matrix form as shown below

$$\begin{bmatrix} 1 & 2 & 3 & 4 \\ 5 & 5 & 7 & 12 \\ 6 & 7 & 10 & 16 \end{bmatrix} \begin{bmatrix} x_1 \\ x_2 \\ x_3 \\ x_4 \end{bmatrix} = \begin{bmatrix} 1 \\ 2 \\ 3 \end{bmatrix}$$

Note that $A = \begin{bmatrix} 1 & 2 & 3 & 4 \\ 5 & 5 & 7 & 12 \\ 6 & 7 & 10 & 16 \end{bmatrix}$ The $\underline{\boldsymbol{x}} = \begin{bmatrix} x1 \\ x2 \\ x3 \\ x4 \end{bmatrix}$ and $\underline{\boldsymbol{b}} = \begin{bmatrix} 1 \\ 2 \\ 3 \end{bmatrix}$

The above equation can be viewed as

$$\begin{bmatrix} 1 \\ 5 \\ 6 \end{bmatrix} x1 + \begin{bmatrix} 2 \\ 5 \\ 7 \end{bmatrix} x2 + \begin{bmatrix} 3 \\ 7 \\ 10 \end{bmatrix} x3 + \begin{bmatrix} 4 \\ 12 \\ 16 \end{bmatrix} x4 = \begin{bmatrix} 1 \\ 2 \\ 3 \end{bmatrix}$$

From the above, we can interpret that the solution for the above linear equation occurs only if the vector $\underline{\boldsymbol{b}}$ lies in the column space of the matrix A.

1.15.1 Trick to Obtain the Solution

$$\begin{bmatrix} 1 & 2 & 3 & 4 \\ 5 & 5 & 7 & 12 \\ 6 & 7 & 10 & 16 \end{bmatrix} \begin{bmatrix} x1 \\ x2 \\ x3 \\ x4 \end{bmatrix} = \begin{bmatrix} 1 \\ 2 \\ 3 \end{bmatrix}$$

$$\Rightarrow \begin{bmatrix} 1 & 0 & 0 \\ 0 & 1 & 0 \\ 0 & 0 & 1 \end{bmatrix} \begin{bmatrix} 1 & 2 & 3 & 4 \\ 5 & 5 & 7 & 12 \\ 6 & 7 & 10 & 16 \end{bmatrix} \begin{bmatrix} x1 \\ x2 \\ x3 \\ x4 \end{bmatrix} = \begin{bmatrix} 1 & 0 & 0 \\ 0 & 1 & 0 \\ 0 & 0 & 1 \end{bmatrix} \begin{bmatrix} 1 \\ 2 \\ 3 \end{bmatrix}$$

Applying the Row operation on the identity matrix on both sides will not affect the equality

$$\mathbf{R2\text{-}>R2\text{-}5*R1}$$

$$\mathbf{R3\text{-}>R3\text{-}6*R1}$$

$$\begin{bmatrix} 1 & 0 & 0 \\ -5 & 1 & 0 \\ -6 & 0 & 1 \end{bmatrix} \begin{bmatrix} 1 & 2 & 3 & 4 \\ 5 & 5 & 7 & 12 \\ 6 & 7 & 10 & 16 \end{bmatrix} \begin{bmatrix} x1 \\ x2 \\ x3 \\ x4 \end{bmatrix} = \begin{bmatrix} 1 & 0 & 0 \\ -5 & 1 & 0 \\ -6 & 0 & 1 \end{bmatrix} \begin{bmatrix} 1 \\ 2 \\ 3 \end{bmatrix}$$

$$\Rightarrow \begin{bmatrix} 1 & 0 & 0 \\ 0 & 1 & 0 \\ 0 & 0 & 1 \end{bmatrix} \begin{bmatrix} 1 & 2 & 3 & 4 \\ 0 & -5 & -8 & -8 \\ 0 & -5 & -8 & -8 \end{bmatrix} \begin{bmatrix} x1 \\ x2 \\ x3 \\ x4 \end{bmatrix} = \begin{bmatrix} 1 & 0 & 0 \\ 0 & 1 & 0 \\ 0 & 0 & 1 \end{bmatrix} \begin{bmatrix} 1 \\ -3 \\ -3 \end{bmatrix}$$

Applying again the Row operation

$$\mathbf{R3\text{-}>R3\text{-}R2}$$

$$\Rightarrow \begin{bmatrix} 1 & 0 & 0 \\ 0 & 1 & 0 \\ 0 & -1 & 1 \end{bmatrix} \begin{bmatrix} 1 & 2 & 3 & 4 \\ 0 & -5 & -8 & -8 \\ 0 & -5 & -8 & -8 \end{bmatrix} \begin{bmatrix} x1 \\ x2 \\ x3 \\ x4 \end{bmatrix} = \begin{bmatrix} 1 & 0 & 0 \\ 0 & 1 & 0 \\ 0 & -1 & 1 \end{bmatrix} \begin{bmatrix} 1 \\ -3 \\ -3 \end{bmatrix}$$

We get

$$\Rightarrow \begin{bmatrix} 1 & 2 & 3 & 4 \\ 0 & -5 & -8 & -8 \\ 0 & 0 & 0 & 0 \end{bmatrix} \begin{bmatrix} x1 \\ x2 \\ x3 \\ x4 \end{bmatrix} = \begin{bmatrix} 1 \\ -3 \\ 0 \end{bmatrix}$$

The modified equation mentioned above can be used to compute the values for the unknown variables $x1, x2, x3$ and $x4$.

To get the solution, the vector $\begin{bmatrix} 1 \\ -3 \\ 0 \end{bmatrix}$ should lie in the column space of the matrix $\begin{bmatrix} \mathbf{1} & 2 & 3 & 4 \\ 0 & \mathbf{-5} & -8 & -8 \\ 0 & 0 & 0 & 0 \end{bmatrix}$. The Bold letters mentioned in the matrix are called pivot elements and the corresponding columns are called pivot columns and the remaining columns are called free variable columns. Hence the column space of the above matrix is obtained as

$$x_1 \begin{bmatrix} 1 \\ 0 \\ 0 \end{bmatrix} + x_2 \begin{bmatrix} 2 \\ -5 \\ 0 \end{bmatrix}$$

Thus for some values x_1 and x_2

$$x_1 \begin{bmatrix} 1 \\ 0 \\ 0 \end{bmatrix} + x_2 \begin{bmatrix} 2 \\ -5 \\ 0 \end{bmatrix} = \begin{bmatrix} 1 \\ -3 \\ 0 \end{bmatrix}$$

$$\Rightarrow x_2 = \frac{3}{5} x_1 = \frac{-1}{5}$$

Thus the particular solution of the equation

$$\begin{bmatrix} 1 & 2 & 3 & 4 \\ 5 & 5 & 7 & 12 \\ 6 & 7 & 10 & 16 \end{bmatrix} \begin{bmatrix} x1 \\ x2 \\ x3 \\ x4 \end{bmatrix} = \begin{bmatrix} 1 \\ 2 \\ 3 \end{bmatrix} \text{ is } \begin{bmatrix} x1 \\ x2 \\ x3 \\ x4 \end{bmatrix} = \begin{bmatrix} -1/5 \\ 3/5 \\ 0 \\ 0 \end{bmatrix}$$

Consider the vector $\begin{bmatrix} n1 \\ n2 \\ n3 \\ n4 \end{bmatrix}$ in the null space of the matrix $A = \begin{bmatrix} 1 & 2 & 3 & 4 \\ 5 & 5 & 7 & 12 \\ 6 & 7 & 10 & 16 \end{bmatrix}$

(i.e.)

$$\begin{bmatrix} 1 & 2 & 3 & 4 \\ 5 & 5 & 7 & 12 \\ 6 & 7 & 10 & 16 \end{bmatrix} \begin{bmatrix} n1 \\ n2 \\ n3 \\ n4 \end{bmatrix} = \begin{bmatrix} 0 \\ 0 \\ 0 \end{bmatrix}$$

Combining the two equation, we get

$$\begin{bmatrix} 1 & 2 & 3 & 4 \\ 5 & 5 & 7 & 12 \\ 6 & 7 & 10 & 16 \end{bmatrix} \begin{bmatrix} x1 + n1 \\ x2 + n2 \\ x3 + n3 \\ x4 + n4 \end{bmatrix} = \begin{bmatrix} 1 + 0 \\ 2 + 0 \\ 3 + 0 \end{bmatrix}$$

$$\Rightarrow \begin{bmatrix} 1 & 2 & 3 & 4 \\ 5 & 5 & 7 & 12 \\ 6 & 7 & 10 & 16 \end{bmatrix} \begin{bmatrix} x1 + n1 \\ x2 + n2 \\ x3 + n3 \\ x4 + n4 \end{bmatrix} = \begin{bmatrix} 1 \\ 2 \\ 3 \end{bmatrix}$$

$\Rightarrow \begin{bmatrix} x1 + n1 \\ x2 + n2 \\ x3 + n3 \\ x4 + n4 \end{bmatrix}$ is the solution of the Linear equation.

where $\begin{bmatrix} x1 \\ x2 \\ x3 \\ x4 \end{bmatrix}$ is the particular solution and $\begin{bmatrix} n1 \\ n2 \\ n3 \\ n4 \end{bmatrix}$ is the particular vector in the Null space of the matrix A

So the complete solution of Linear Equation of the form $\mathbf{A}\underline{\boldsymbol{x}} = \underline{\boldsymbol{b}}$ consists of particular solution + any vector in the null space of the matrix A.

Computation of Null space of the matrix A

The matrix A in RREF is given below

$$\begin{bmatrix} \mathbf{1} & 2 & 3 & 4 \\ 0 & \mathbf{-5} & -8 & -8 \\ 0 & 0 & 0 & 0 \end{bmatrix}$$

Representing pivot variables in terms of free variables, we get

$$\begin{aligned} x_2 &= -\frac{8}{5}x_3 + \left(-\frac{8}{5}\right)x_4 \\ x_1 &= -2x_2 - 3x_3 - 4x_4 \\ \Rightarrow x_1 &= -2^*\left(-\frac{8}{5}x_3 + \left(-\frac{8}{5}\right)x_4\right) - 3x_3 - 4x_4 \\ \Rightarrow x_1 &= \frac{16}{5}x_3 + \frac{16}{5}x_4 - 3x_3 - 4x_4 \\ \Rightarrow x_1 &= \frac{1}{5}x_3 + \frac{-4}{5}x_4 \end{aligned}$$

The Null space of the matrix A is given as

$$\begin{bmatrix} \frac{1}{5}x3 - \frac{4}{5}x4 \\ -\frac{8}{5}x3 - \frac{8}{5}x4 \\ x3 \\ x4 \end{bmatrix} \Rightarrow x3\begin{bmatrix} \frac{1}{5} \\ -\frac{8}{5} \\ 1 \\ 0 \end{bmatrix} + x4\begin{bmatrix} -\frac{4}{5} \\ -\frac{8}{5} \\ 0 \\ 1 \end{bmatrix}$$

Thus the Complete solution of the Linear Equation is given as

$$\begin{bmatrix} -\frac{1}{5} \\ \frac{3}{5} \\ 0 \\ 0 \end{bmatrix} + x3\begin{bmatrix} \frac{1}{5} \\ -\frac{8}{5} \\ 1 \\ 0 \end{bmatrix} + x4\begin{bmatrix} -\frac{4}{5} \\ -\frac{8}{5} \\ 0 \\ 1 \end{bmatrix}, \text{ where } x3, x4 \in \mathbb{R}$$

Case 2: If the vector $\underline{b}$ does not lie in the column space of the matrix A

For any $\underline{x} \in \mathbb{R}^m$, $A\underline{x}$ always lies in the column space of the matrix A and hence If the vector $\underline{\boldsymbol{b}}$ does not lie in the column space of the matrix A, then solution doesn't occur.

In this case the vector $\underline{x}$ is estimated such that $\|A\underline{x} - \underline{\boldsymbol{b}}\|$ is minimized. It is also known that the solution occurs in the vector space $\mathbb{R}^m$ (i.e.) $\underline{x} \in \mathbb{R}^m$ Any vector in the vector space $\mathbb{R}^m$ can be represented as the direct sum of the vector from the column space and the vector from the Left Null space of the matrix A. The vector b_c that lies in the column space of the matrix A are to be found such that $\|b_c - \underline{\boldsymbol{b}}\|$ is minimized.

Note:

$$\text{The vector } \underline{b} = b_c + b_{ln}.$$

Multiplying the matrix

$$\begin{aligned}
&A^T \text{on both sides, we get}\\
&A^T\underline{\boldsymbol{b}} = A^T(b_c + b_{ln})\\
&= A^T b_c + A^T b_{ln}\\
&= A^T b_c + 0 \text{ [Because } b_{ln} \text{ lies in the left Null Space of the matrix A]}\\
&\Rightarrow A^T\underline{\boldsymbol{b}} = A^T b_c
\end{aligned}$$

Consider solving the equation $A\underline{x} = \underline{\boldsymbol{b}}$ when $\underline{\boldsymbol{b}}$ does not lie in the column space of the matrix A.

The vector $\underline{x}$ cannot be found, because $\underline{\boldsymbol{b}}$ does not lie in the column space of A. Hence the best value for the vector $\underline{x}$ is estimated as $\widehat{\underline{x}}$ such that

$$A\widehat{\underline{x}} = \underline{b_c}$$

Multiplying the matrix A^T on both sides, we get

$$\begin{aligned}
&A^T A\widehat{\underline{x}} = A^T\underline{b_c}\\
&\Rightarrow \widehat{\underline{x}} = (A^T A)^{-1} A^T\, b_c\\
&\Rightarrow \widehat{\underline{x}} = (A^T A)^{-1} A^T\underline{\boldsymbol{b}} \qquad [\because A^T\, b_c = A^T\, \underline{\boldsymbol{b}}]
\end{aligned}$$

Note that $(A^T A)^{-1}(A^T A)$ is the identity matrix and hence $(A^T A)^{-1} A^T$ is the left inverse of the matrix A.

Also note that the best estimate $\widehat{\underline{x}} = (A^T A)^{-1} A^T\ \underline{\boldsymbol{b}}$ exists only when $A^T A$ is invertible. Also note that $A^T A$ is invertible if and only if the columns of A are independent (i.e.) A is invertible. This is true because Null space of A is exactly equal to the Null space of $A^T A$. So if $N(A)$ is $\underline{0}$, then $N(A^T A)$ is $\underline{0}$ and hence if A is invertible, $A^T A$ is invertible and vice versa.

It can also be shown that the above estimated value for $\underline{x}$ (i.e.) $\widehat{\underline{x}}$ is the estimate of the Multi variable $\underline{x}$ such that $\|A\underline{x} - \underline{\boldsymbol{b}}\|$ is minimized.

Example 1.17. Solving the Linear Equation of the form $A\underline{x} = \underline{\boldsymbol{b}}$

$$\text{Where } A = \begin{bmatrix} 1 & 4 \\ 2 & 5 \\ 3 & 6 \end{bmatrix} \underline{b} = \begin{bmatrix} 5 \\ 7 \\ 10 \end{bmatrix}$$

The best estimate for the variable $\underline{x}$ lies in the vector space $\mathbb{R}^2$.

It is known that

$$\dim(N(A)) + \dim(R(A)) = 2$$
$$\dim(C(A)) + \dim(N(A^T)) = 3$$
$$\dim(N(A)) + \dim(C(A)) = 2$$

Note that the dimension of the column space of the matrix A is 2. From the above mentioned fact, dimension of the Null space of the matrix A is 0, the dimension of the row space of the matrix A is 2 and the dimension of the Left Null space of the matrix A is 1.

The vector $\underline{b}$ does not lie in the column space of A. The best estimate for the variable $\underline{x}$ is found as $\widehat{\underline{x}}$ such that $A\widehat{\underline{x}} = b_c$ and $\|b_c - \underline{\boldsymbol{b}}\|$ is minimized. The vector b_c lies in the column space of the matrix A, which can be represented as

$$b_c = \begin{bmatrix} 1 & 4 \\ 2 & 5 \\ 3 & 6 \end{bmatrix} \begin{bmatrix} \widehat{\underline{x1}} \\ \widehat{\underline{x2}} \end{bmatrix} = A\widehat{\underline{x}}$$

Consider the vectors b_c and $\underline{\boldsymbol{b}}$ as shown below (Fig. 1.3).

From the figure, it can be realized that the distance between the vector b_c and $\underline{b}$ occur only when the vector b_c is orthogonal to the vector $(\underline{b} - b_c)$. As b_c lies in the column space of A, the vector $(\underline{b} - b_c)$ orthogonal to the individual basis of the column space of the matrix A,

$$\Rightarrow \begin{bmatrix} 1 \\ 2 \\ 3 \end{bmatrix}^T \left(\underline{b} - \begin{bmatrix} 1 & 4 \\ 2 & 5 \\ 3 & 6 \end{bmatrix} \begin{bmatrix} \widehat{\underline{x1}} \\ \widehat{\underline{x2}} \end{bmatrix} \right) = \underline{0} \text{- -(1)}$$

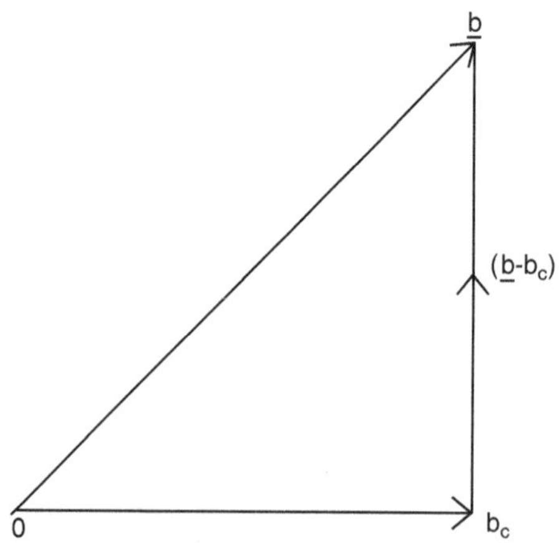

Fig. 1.3 Pictorial representation of the vectors b_c and $\underline{\boldsymbol{b}}$

$$\begin{bmatrix}4\\5\\6\end{bmatrix}^T \left(\underline{b} - \begin{bmatrix}1 & 4\\2 & 5\\3 & 6\end{bmatrix}\begin{bmatrix}\widehat{\underline{x}1}\\ \widehat{\underline{x}2}\end{bmatrix}\right) = \underline{0} \text{- -(2)}$$

Combining both the equations we get

$$\begin{bmatrix}1 & 4\\2 & 5\\3 & 6\end{bmatrix}^T \left(\underline{b} - \begin{bmatrix}1 & 4\\2 & 5\\3 & 6\end{bmatrix}\begin{bmatrix}\widehat{\underline{x}1}\\ \widehat{\underline{x}2}\end{bmatrix}\right) = \underline{0}$$

$$\Rightarrow \begin{bmatrix}1 & 4\\2 & 5\\3 & 6\end{bmatrix}^T \underline{b} = \begin{bmatrix}1 & 4\\2 & 5\\3 & 6\end{bmatrix}^T \begin{bmatrix}1 & 4\\2 & 5\\3 & 6\end{bmatrix}\begin{bmatrix}\widehat{\underline{x}1}\\ \widehat{\underline{x}2}\end{bmatrix}$$

$$\Rightarrow \begin{bmatrix}1 & 4\\2 & 5\\3 & 6\end{bmatrix}^T \begin{bmatrix}1 & 4\\2 & 5\\3 & 6\end{bmatrix}\begin{bmatrix}\widehat{\underline{x}1}\\ \widehat{\underline{x}2}\end{bmatrix} = \begin{bmatrix}1 & 4\\2 & 5\\3 & 6\end{bmatrix}^T \underline{b}$$

$$\Rightarrow \begin{bmatrix}\widehat{\underline{x}1}\\ \widehat{\underline{x}2}\end{bmatrix} = \left(\begin{bmatrix}1 & 4\\2 & 5\\3 & 6\end{bmatrix}^T \begin{bmatrix}1 & 4\\2 & 5\\3 & 6\end{bmatrix}\right)^{-1} \begin{bmatrix}1 & 4\\2 & 5\\3 & 6\end{bmatrix}^T \underline{b}$$

$$\Rightarrow \begin{bmatrix}\widehat{\underline{x}1}\\ \widehat{\underline{x}2}\end{bmatrix} = \left(\begin{bmatrix}1 & 4\\2 & 5\\3 & 6\end{bmatrix}^T \begin{bmatrix}1 & 4\\2 & 5\\3 & 6\end{bmatrix}\right)^{-1} \begin{bmatrix}1 & 4\\2 & 5\\3 & 6\end{bmatrix}^T \begin{bmatrix}5\\7\\10\end{bmatrix} = [\widehat{\underline{x}}] = (A^T A)^{-1} A^T \underline{\boldsymbol{b}}$$

which is same as the one used in the previous section.

$$[\widehat{\underline{x}}] = \begin{bmatrix}1.7222\\0.7778\end{bmatrix}$$

$$[\widehat{\underline{x}}] = (A^T A)^{-1} A^T \underline{\boldsymbol{b}}$$

Multiplying both sides by A

$$A[\widehat{\underline{x}}] = (A^T A)^{-1} A^T \underline{\boldsymbol{b}}$$

$A[\widehat{\underline{x}}]$ is the vector b_c that lies in the column space of the matrix A.

$$\Rightarrow A(A^T A)^{-1} A^T \underline{\boldsymbol{b}} = b_c$$

$A(A^T\ A)^{-1}\ A^T$ is called as the **projection matrix**, that projects the vector in the vector space $\mathbb{R}^3$ into the column space of the matrix A such that $\|b_c - \underline{b}\|$ is minimized.

In this example Projection matrix is

$$\begin{bmatrix} 0.8333 & 0.3333 & -0.1667 \\ 0.3333 & 0.3333 & 0.3333 \\ -0.1667 & 0.3333 & 0.8333 \end{bmatrix}$$

$(A^T\ A)^{-1}(A^T\text{A})$ = Identity matrix. Hence it is also clear that $(A^T\ A)^{-1}\ A^T$ is the left inverse of the matrix A, which is usually represented as A^+.

In this example Left inverse matrix of the matrix A is given as

$$\begin{bmatrix} -0.9444 & -0.1111 & 0.7222 \\ 0.4444 & 0.1111 & -0.2222 \end{bmatrix}$$

$$\begin{bmatrix} -0.9444 & -0.1111 & 0.7222 \\ 0.4444 & 0.1111 & -0.2222 \end{bmatrix} \begin{bmatrix} 1 & 4 \\ 2 & 5 \\ 3 & 6 \end{bmatrix} = \begin{bmatrix} 1 & 0 \\ 0 & 1 \end{bmatrix}$$

Left inverse of the matrix exists if $(A^T\ A)^{-1}$ exists. Similarly if $(AA^T)\ (AA^T)^{-1}$ = Identity matrix, $A^T\ (AA^T)^{-1}$ is the right inverse of the matrix A. Right inverse of the matrix exists only when $(AA^T)^{-1}$ exists.

As already mentioned, if the matrix A consists of all columns independent, $(A^T\ A)^{-1}$ exists and hence Left inverse of the matrix exists. Similarly if all the rows of the matrix A are independent (i.e.) all the columns of the matrix A^T is independent, Right inverse of the matrix exists.

Thus solving the equation $A\underline{x} = \underline{\boldsymbol{b}}$ when $\underline{\boldsymbol{b}}$ is not in the column space of the matrix A (the matrix with all the columns are independent to each other), can be obtained by multiplying the left inverse matrix of the matrix A represented as A^+ with the vector $\underline{\boldsymbol{b}}$

$$\boxed{\hat{x}} = (A^T A)^{-1} A^T\ \underline{\boldsymbol{b}} = A^+ b$$

Note: Solving the equation $A\underline{x} = \underline{b}$, when $\underline{b}$ is not in the column space of the matrix A and $(A^T\ A)^{-1}$ doesn't exist can be obtained using Singular Value Decomposition (SVD) which is referred in the Section 4.25.

1.16 Gram Schmidt Orthonormalization Procedure for Obtaining Orthonormal Basis

1. Given the set of independent vectors {v1, v2, v3, ... vn} which forms the basis of the vector space V, an alternative orthonormal basis {a1, a2, ... an} for the vector space 'V' can be obtained using Gram-Schmidt Orthonormalization procedure as described below (Fig. 1.4).

   Steps

1. $a1 = \frac{v1}{\|v1\|}$

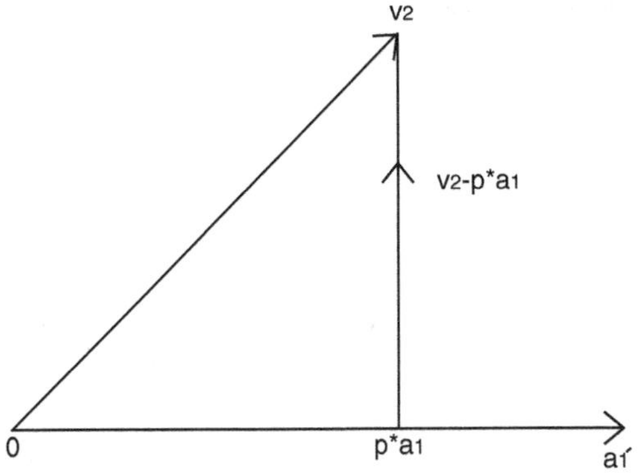

Fig. 1.4 Illustration of Gram-Schmidt orthogonalization

Find out the vector $a2$ corresponding to the vector $v1$ that is orthogonal to the vector $a1$

2. The vector that is orthogonal to the vector $a1$ is $v2 - p * a1$ as shown in the figure, where $p * a1$ is the perpendicular projection of the vector v2 on the vector $a1$. The projection point p is computed using the following condition.

$$(v2 - pa1)^T a1 =$$
$$\Rightarrow (v2)^T a1 = (pa1)^T a1$$
$$\Rightarrow \mathrm{p} = \frac{v2^T a1}{a1^T a1} = v2^T a1 [\because\ a1^T a1 = 1] = a1^T v2$$

Thus the orthogonal vector q2 to the vector $a1$ is obtained as $\mathrm{q2} = v2 - (a1^T\ v2) * a1$ and the corresponding orthonormal vector is given as $\mathrm{a2} = \frac{q2}{\|q2\|}$

3. Now we are in need of the vector which is orthogonal to both $a1$ and $a2$ corresponding to the vector $v3$

   The perpendicular projection vector of the vector $v3$ on the column space of the matrix consists of a1 and a2 as their columns is obtained as $p1 * a1 + p2 * a2$ (as shown in Fig. 1.5). The vector $v3 - p1 * a1 + p2 * a2$ is perpendicular to both the vectors a1 and a2. The projection points $p1, p2$ are obtained as the perpendicular projection of the vector $v3$ on the vector $a1$ and $a2$ respectively which are computed as

$$\mathrm{p1} = \frac{v3^T a1}{a1^T a1} = v3^T a1 [\because a1^T a1 = 1] = a1^T v3$$
$$\mathrm{p2} = \frac{v3^T a2}{a2^T a2} = v3^T a2 [\because a2^T a2 = 1] = a2^T v3$$

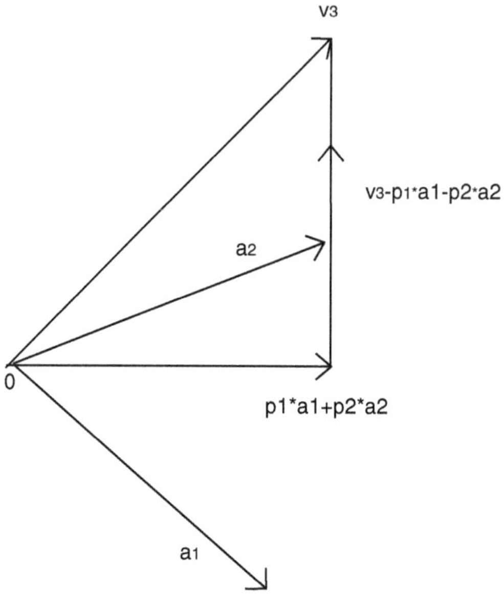

Fig. 1.5 Illustration of Gram-Schmidt orthogonalization

Thus the vector perpendicular to both the orthonormal vectors $a1$ *and* $a2$ is given as $q3 = v3 - a1^T\ v3 * a1 - a2^T\ v3 * a2$ -(3) and the corresponding orthonormal vector which is perpendicular to both $a1$ *and* $a2$ is given as

$$a3 = \frac{q3}{\|q3\|}$$

This can also be computed using the Projection matrix as follows.
The perpendicular projection vector $q3'$ of the vector of the vector $v3$ on the column space of the matrix consists of a1 and a2 as their columns is obtained using projection matrix as follows
Let A $= [a1\ a2]$ (Note that $a1$ *and* $a2$ are the column vectors).

$$\begin{aligned}
\text{Projection vector } q3' &= A(A^T A)^{-1} A^T\ q3 \\
&= \left([a1\ a2] \left(\begin{bmatrix} a1^T \\ a2^T \end{bmatrix} [a1\ a2] \right)^{-1} \begin{bmatrix} a1^T \\ a2^T \end{bmatrix} \right) q3 \\
&= \left([a1\ a2]\, (I)^{-1} \begin{bmatrix} a1^T \\ a2^T \end{bmatrix} \right) q3 \\
&= \left([a1\ a2] \begin{bmatrix} a1^T \\ a2^T \end{bmatrix} \right) q3 \\
&= (a1a1^T + a2^T a2) q3
\end{aligned}$$

Therefore perpendicular vector $q3 = v3 - q3' = v3 - (a1a1^T + a2^T\ a2)q3$ that is same as Eq. 3 as mentioned above.

4. Similarly the vector perpendicular to both the orthonormal vectors $a1$ *and* $a2,\ a3,\ a4, \ldots an-1$ corresponding to the independent vector vn is given as

$$qn = vn - a1^T vn * a1 - a2^T vn * a2 - a3^T vn * a3$$

$-a4^T\ vn * a4 - \cdots a_{n-1}{}^T vn \quad {}^*a_{n-1}$ and the corresponding orthonormal vector is

$$an = \frac{qn}{\|qn\|}$$

Thus the set of orthonormal basis $\{a1, a2, \ldots an\}$, of the vector space 'V' corresponding to the basis vectors $\{v1, v2, \ldots, vn\}$ is obtained using Gram-Schmidt orthogonalization procedure.

Example 1.18. Consider the matrix A $= \begin{bmatrix} 1 & 5 & 10 \\ 2 & 6 & 10 \\ 3 & 7 & 11 \\ 4 & 8 & 12 \end{bmatrix}$ whose column vectors are independent to each other.

Using Gram-Schmidt, the corresponding orthonormal vectors are obtained as follows

Let $v1 = \begin{bmatrix} 1 \\ 2 \\ 3 \\ 4 \end{bmatrix}$, $v2 = \begin{bmatrix} 5 \\ 6 \\ 7 \\ 8 \end{bmatrix}$ and $v3 = \begin{bmatrix} 10 \\ 10 \\ 11 \\ 12 \end{bmatrix}$

$$a1 = \frac{v1}{\|v1\|} = \begin{bmatrix} 0.1826 \\ 0.3651 \\ 0.5477 \\ 0.7303 \end{bmatrix}$$

$$q2 = v2 - (a1^T v2) * a1 = \begin{bmatrix} 2.6667 \\ 1.3333 \\ 0.0000 \\ -1.3333 \end{bmatrix}$$

$$a2 = \frac{q2}{\|q2\|} = \begin{bmatrix} 0.1865 \\ 0.4082 \\ 0.0000 \\ -0.4082 \end{bmatrix}$$

$$q3 = v3 - a1^T v3 * a1 - a2^T v3 * a2 = \begin{bmatrix} 0.3000 \\ -0.4000 \\ -0.1000 \\ 0.2000 \end{bmatrix}$$

$$a3 = \frac{q3}{\|q3\|} = \begin{bmatrix} 0.5477 \\ -0.7303 \\ -0.1826 \\ 0.3651 \end{bmatrix}$$

Thus the set of orthonormal vectors are

$$\begin{bmatrix} 0.1826 & 0.8165 & 0.5477 \\ 0.3651 & 0.4082 & -0.7303 \\ 0.5477 & 0.0000 & -0.1826 \\ 0.7303 & -0.4082 & 0.3651 \end{bmatrix}$$

1.17 QR Factorization

The matrix with independent column vectors can be represented as the product of the matrix with orthonormal column vectors and the upper triangular matrix. Consider the matrix A as shown below.

$A = [v1\ v2\ \ldots\ vn]$, where $v1\ v2\ \ldots\ vn$ are the independent vectors.

From Gram Schmidt orthogonalization procedure discussed in the previous section,

$$a1 = \frac{v1}{\|v1\|}$$
$$\Rightarrow v1 = a1\|v1\|$$

Multiplying $a1^T$ on both sides, we get

$$\Rightarrow a1^T v1 = a1^T a1\|v1\|$$
$$\Rightarrow \|v1\| = a1^T v1$$

From the above, we get $v1 = a1(a1^T\ v1)$

Similarly

$$\text{a2} = \frac{q2}{\|q2\|}$$
$$\Rightarrow \|q2\|a2 = q2$$

Multiplying $a2^T$ on both sides, we get

$$\Rightarrow a2^T\ q2 = a2^T a2\|q2\|$$
$$\Rightarrow \|q2\| = a2^T\ q2$$

Also

$$q2 = v2 - (a1^T v2) * a1 \text{ (see previous section)}$$

Multiplying $a2^T$ on both sides, we get

$$a2^T q2 = a2^T v2 - a2^T * (a1^T v2) * a1$$
$$\Rightarrow a2^T q2 = a2^T v2$$

Also we know

$$\|q2\| = a2^T q2$$
$$\Rightarrow a2^T v2 = \|q2\|$$

Rewriting the equation for $v2$ we get,

$$v2 = q2 + (a1^T v2) * a1$$
$$\Rightarrow v2 = \|q2\| a2 + (a1^T v2) * a1$$
$$\Rightarrow v2 = (a2^T v2) * a2 + (a1^T v2) * a1$$

Similarly it can be shown that

$$\Rightarrow v3 = a3 * (a3^T v3) + a2 * (a2^T v3) + a1 * (a1^T v3)$$

Similarly it can be shown that

$$\Rightarrow vn = an * (an^T\ vn) + a_{n-1} * (a_{n-1}{}^T vn) + a_{n-2} * (a_{n-2}{}^T vn)$$
$$+ a_{n-3} * (a_{n-3}{}^T vn) + a_{n-4} * (a_{n-4}{}^T vn) + \cdots a_1 * (a_1{}^T vn)$$

Representing the above list of equations for $v1, v2, v3, \ldots vn$ in matrix form

$$A = \begin{bmatrix} v1 & v2 & v3 & \ldots & vn \end{bmatrix}$$

$$= \begin{bmatrix} a1\ a2\ a3 \ldots an \end{bmatrix} \begin{bmatrix} (a1^T v1) & (a1^T v2) & (a1^T v3) & \cdots & (a1^T vn) \\ 0 & (a2^T v2) & (a2^T v3) & \cdots & (a2^T vn) \\ 0 & 0 & (a3^T v3) & \cdots & (a3^T vn) \\ 0 & 0 & 0 & \cdots & (a4^T vn) \\ 0 & 0 & 0 & \cdots & (a5^T vn) \\ 0 & 0 & 0 & \cdots & (a6^T vn) \\ \ldots & \ldots & \ldots & \cdots & \ldots \\ 0 & 0 & 0 & \cdots & (an^T vn) \end{bmatrix}$$

Thus the matrix A is represented as the product of the matrix consists of orthonormal vectors and the upper triangular matrix as shown above.

Example 1.19. From the Example 1.17, QR factorization of the matrix A is given below

$$\begin{bmatrix} 1 & 5 & 10 \\ 2 & 6 & 10 \\ 3 & 7 & 11 \\ 4 & 8 & 12 \end{bmatrix} =$$

$$\begin{bmatrix} 0.1826 & 0.8165 & 0.5477 \\ 0.3651 & 0.4082 & -0.7303 \\ 0.5477 & 0.0000 & -0.1826 \\ 0.7303 & -0.4082 & 0.3651 \end{bmatrix} \begin{bmatrix} 5.4772 & 12.7802 & 20.2657 \\ 0 & 3.2660 & 7.3485 \\ 0 & 0 & i0.5477 \end{bmatrix}$$

1.18 Eigen Values and Eigen Vectors

For the given matrix A, if there exists the non zero vector $\underline{x}$ such that $A\underline{x} = \lambda\underline{x}$, where λ is the scaling factor, then the vector $\underline{x}$ is the Eigen vector corresponding to the matrix A. $\lambda\underline{x}$ is in the column space of the matrix A. As λ is the scalar, **the Eigen vector $\underline{x}$ lies in the column space of the matrix A**.

$$\Rightarrow [A - \lambda I]\underline{x} = \underline{0}$$
$$\textit{where } I \textit{ is the Identity matrix}$$

$\Rightarrow$ The Null space of the matrix $[A - \lambda I] \neq \underline{0}$
$\Rightarrow$ If the matrix A is the square matrix, the matrix $[A - \lambda I]$ is called as singular matrix.
$\Rightarrow$ Determinant $(A - \lambda I) = 0$, which is represented as $|A - \lambda I| = 0$ is the polynomial equation with variable λ. The equation thus obtained is called characteristic equation and the solutions to the characteristic equation are called Eigen values. Let the degree of the characteristic polynomial be 'n' and the corresponding Eigen values are $\lambda 1, \lambda 2, \lambda 3, \ldots \lambda n$ (say). Eigen values thus obtained need not be distinct.

For every distinct Eigen values, there exist one or more Eigen vectors that are obtained as follows. The Eigen vectors corresponding to the Eigen value λ_1 are obtained as the basis of the Null space of the matrix $[A - \lambda_1 I]$.

Example 1.20. Consider the matrix A = $\begin{bmatrix} 1 & 0 & 0 \\ 0 & 2 & 3 \\ 0 & 4 & 5 \end{bmatrix}$

The Characteristic polynomial of the matrix A is given as

$$-\lambda^3 + 8\lambda^2 - 5\lambda - 2 = 0$$

The Eigen values of the matrix A are given as

$$\lambda_1 = 1$$
$$\lambda_2 = \frac{7 + \sqrt{57}}{2} = 7.2749$$
$$\lambda_3 = \frac{7 - \sqrt{57}}{2} = -0.2749$$

The Eigen vectors corresponding to the distinct Eigen values are obtained as follows. Eigen vector corresponding to the Eigen value $\lambda_1 = 1$ is the null space of the matrix

$$[A - \lambda_1 I] = A = \begin{bmatrix} 1 & 0 & 0 \\ 0 & 2 & 3 \\ 0 & 4 & 5 \end{bmatrix} - 1 \begin{bmatrix} 1 & 0 & 0 \\ 0 & 1 & 0 \\ 0 & 0 & 1 \end{bmatrix} = \begin{bmatrix} 0 & 0 & 0 \\ 0 & 1 & 3 \\ 0 & 4 & 4 \end{bmatrix}$$

The Null space of the matrix $\begin{bmatrix} 0 & 0 & 0 \\ 0 & 1 & 3 \\ 0 & 4 & 4 \end{bmatrix}$ is obtained as follows.

$$\begin{bmatrix} 1 \\ 0 \\ 0 \end{bmatrix} x, x, \in \mathbb{R}$$

Similarly the Null space of the matrix $[A - \lambda_2 I]$ is obtained as

$$\begin{bmatrix} 0 \\ 0.4944 \\ 0.8693 \end{bmatrix} x, x \in \mathbb{R}$$

and the Null space of the matrix $[A - \lambda_3 I]$ is obtained as

$$\begin{bmatrix} 0 \\ -0.7968 \\ 0.6042 \end{bmatrix} x, x \in \mathbb{R}$$

Thus the Eigen vectors are $\begin{bmatrix} 1 \\ 0 \\ 0 \end{bmatrix}, \begin{bmatrix} 0 \\ 0.4944 \\ 0.8693 \end{bmatrix}, \begin{bmatrix} 0 \\ -0.7968 \\ 0.6042 \end{bmatrix}$

1.19 Geometric Multiplicity (Versus) Algebraic Multiplicity

Let the distinct Eigen values of the arbitrary matrix 'A' are $\lambda_1, \lambda_2, \lambda_3, \ldots \lambda_k$ such that λ_1 is repeated n_1 times, λ_2 is repeated n_2 times and similarly λ_k is repeated n_k times. The sequence of numbers $n_1, n_2, n_3, \ldots n_k$ are called Algebraic multiplicity of the corresponding Eigen values $\lambda_1, \lambda_2, \lambda_3, \ldots \lambda_k$. The dimension of the Null space of the following matrices $[A - \lambda_1 I], [A - \lambda_2 I], [A - \lambda_3 I], \ldots [A - \lambda_k I]$ are $m_1, m_2, m_3, \ldots m_k$ respectively. The sequence of numbers $m_1, m_2, m_3, \ldots, m_k$ are called Geometric multiplicity corresponding to the Eigen values $\lambda_1, \lambda_2, \lambda_3, \ldots \lambda_k$.

For any Eigen value 'λ_k', $m_k \leq n_k$. If $m_k < n_k$, the concern matrix 'A' is called deficient matrix.

Note:

1. Eigen vectors corresponding to distinct Eigen values are independent.

   Proof. Let $\lambda_1, \lambda_2, \lambda_3$ forms the Eigen values of the matrix A and the corresponding Eigen vectors are e_1, e_2, e_3 respectively.

$$\text{Let suppose } c_1 e_1 + c_2 e_2 + c_3 e_3 = 0 - - - - - - - - - - - - - (a).$$

Multiply the matrix A on both sides, we get

$$c_1 Ae_1 + c_2 Ae_2 + c_3 Ae_3 0.$$
$$\Rightarrow c_1 Ae_1 + c_2 Ae_2 + c_3 Ae_3 = 0.$$
$$\Rightarrow c_1 \lambda_1 e_1 + c_2 \lambda_2 e_2 + c_3 \lambda_3 e_3 = 0 - - - - - - - - - - - - (b).$$
$$(b) - \lambda_1 (a)$$
$$we\ get$$
$$c_l(\lambda_1 - \lambda_3)e_1 + c_2(\lambda_2 - \lambda_3)e_2 = 0 - - - - - - - - - - - - (c)$$

Multiplying the matrix A on both sides, we get

$$c_1(\lambda_1 - \lambda_3)A\, e_1 + c_2(\lambda_2 - \lambda_3)Ae_2 = 0$$
$$\Rightarrow c_1(\lambda_1 - \lambda_3)\lambda_1 e_1 + c_2(\lambda_2 - \lambda_3)\lambda_2 e_2$$
$$= 0 - - - - - - - - - (d)$$
$$(d) - \lambda_2(c)$$
$$we\ get$$
$$c_1(\lambda_1 - \lambda_3)(\lambda_1 - \lambda_2 = 0$$
$$\Rightarrow c_1 = 0\, [\because\ \lambda_1, \lambda_2, \lambda_3\ are\ distinct\ Eigen\ values]$$

Similarly it can be shown that $c_2 = 0$ *and* $c_3 = 0$
Hence proved (See Section 1.6)

In the Example 1.19, the Eigen vectors corresponding to distinct Eigen values $1, 7.2749$ and -0.2749 are $\begin{bmatrix} 1 \\ 0 \\ 0 \end{bmatrix}, \begin{bmatrix} 0 \\ 0.4944 \\ 0.8693 \end{bmatrix}, \begin{bmatrix} 0 \\ -0.7968 \\ 0.6042 \end{bmatrix}$. They are independent vectors.

2. Note that Eigen vectors corresponding to the particular Eigen value is the basis of the null space matrix of the form $[A - \lambda_1 I]$, *where* λ_1 is the Eigen value of the matrix A and hence they are independent vectors. Also Eigen vectors corresponding to distinct Eigen values are also independent. Number of Eigen values with or without repetition is equal to the degree of the characteristic function. Thus if the matrix A of size m × n is not deficient, Eigen vectors of the matrix A forms the basis for the $\mathbb{R}^m$.
In the Example 1.19, the matrix A of size 3×3 is not the deficient matrix because of the following reasons.
   - Geometric multiplicity of the Eigen value 1 = Algebraic multiplicity of the particular Eigen value 1 = 1
   - Geometric multiplicity of the Eigen value 7.2749 = Algebraic multiplicity of the particular Eigen value 7.2749 = 1
   - Geometric multiplicity of the Eigen value –0.2749 = Algebraic multiplicity of the particular Eigen value –0.2749 = 1

   Thus the Eigen vectors mentioned above forms the basis of the vector space $\mathbb{R}^3$
3. Let Determinant

$$([A - \lambda I] = (\lambda - \lambda_1)(\lambda - \lambda_2)(\lambda - \lambda_3)\ldots(\lambda - \lambda_n)$$

   Substituting the value for $\lambda = 0$ on both sides, we get
   Determinant

$$([A]) = (0 - \lambda_1)(0 - \lambda_2)(0 - \lambda_3)\ldots(0 - \lambda_n) =$$
$$= (-1)^n \lambda_1 \lambda_2 \lambda_3 \ldots \lambda_n$$

   It can also be shown that trace of the matrix $= a_1 + a_2 + a_3 + \cdots a_n$ $= \lambda_1 + \lambda_2 + \lambda_3 + \cdots + \lambda_n$ by comparing the co-efficient of $(-\lambda)^{n-1}$
4. Let one of the similar matrix for the arbitrary matrix A of size n × n be V and is obtained using the arbitrary invertible matrix M of size n × n as B = $M^{-1}AM$. It can be shown that the Eigen values of the matrices A and B are equal. Also it can also be shown that if x is the Eigen vector of the matrix A, then $M^{-1}x$ is the Eigen vector of the matrix B.

   Proof. Consider the characteristic of the matrix B as $|B - \lambda I| = 0|$. Substitute B = $M^{-1}AM$ in the equation we get $|M^{-1}AM - \lambda I| = 0|$

$$\Rightarrow |M^{-1}AM - \lambda M^{-1}M| = 0|$$
$$\Rightarrow |M^{-1}(A - \lambda I)M| = 0|$$

$\Rightarrow |(A - \lambda I)| = 0|$

$\Rightarrow$ The Characteristic equation of the matrix A and the Characteristic equation of the matrix B are the same and hence their Eigen values are equal. If 'x' is the Eigen vector of the matrix A corresponding to the Eigen value 'λ'. $Ax = \lambda x. \Rightarrow MBM^{-1}x = \lambda x \Rightarrow B(M^{-1}x) = \lambda(M^{-1}x)$

Thus $M^{-1}x$ is the Eigen vector of the matrix B.

5. Computing Eigen vector for the block diagonal matrix.
   Consider the matrix of the form

$$C = \begin{bmatrix} 1 & 2 & 3 & 0 & 0 & 0 \\ 4 & 5 & 6 & 0 & 0 & 0 \\ 7 & 8 & 9 & 0 & 0 & 0 \\ 0 & 0 & 0 & 10 & 11 & 12 \\ 0 & 0 & 0 & 13 & 14 & 15 \\ 0 & 0 & 0 & 16 & 17 & 18 \end{bmatrix} = \begin{bmatrix} A & 0 \\ 0 & B \end{bmatrix}, \text{ where}$$

$$A = \begin{bmatrix} 1 & 2 & 3 \\ 4 & 5 & 6 \\ 7 & 8 & 9 \end{bmatrix} \quad B = \begin{bmatrix} 10 & 11 & 12 \\ 13 & 14 & 15 \\ 16 & 17 & 18 \end{bmatrix}$$

Eigen vectors of the matrix A (Arranged column wise) are given as

$$\begin{bmatrix} -0.2320 & -0.7858 & 0.4082 \\ -0.5253 & -0.0868 & -0.8165 \\ -0.8187 & 0.6123 & 0.4082 \end{bmatrix}$$

Similarly Eigen vectors of the matrix B (Arranged column wise) are given as

$$\begin{bmatrix} -0.4482 & 0.7392 & 0.4082 \\ -0.5689 & -0.0333 & -0.8165 \\ -0.6896 & 0.6727 & 0.4082 \end{bmatrix}$$

The Eigen vectors of the matrix C are obtained as follows. (See Eigen vectors of A and B.)

$$\begin{bmatrix} -0.2320 & -0.7858 & 0.4082 & 0 & 0 & 0 \\ -0.5253 & -0.0868 & -0.8165 & 0 & 0 & 0 \\ -0.8187 & 0.6123 & 0.4082 & 0 & 0 & 0 \\ 0 & 0 & 0 & -0.4482 & -0.7392 & 0.4082 \\ 0 & 0 & 0 & -0.5689 & -0.0333 & -0.8165 \\ 0 & 0 & 0 & -0.6896 & 0.6727 & 0.4082 \end{bmatrix}$$

1.20 Diagonalization of the Matrix

If the matrix A is not the deficient matrix, it can be diagonalizable as described below.

If the matrix A of size m × n is not the deficient matrix, then there exists 'n' Eigen vectors $e_1, e_2, e_3, e_4, e_5, e_6, \ldots e_n$, corresponding to 'n' Eigen values $\lambda_1\ \lambda_2\ \lambda_3\ \lambda_4\ \lambda_5 \ldots \lambda_n$ with or without repetition, that satisfies the following conditions.

$$\begin{aligned} A\,e_1 &= \lambda_1\,e_1 \\ A\,e_2 &= \lambda_2\,e_2 \\ A\,e_3 &= \lambda_3\,e_3 \\ &\ldots \\ A\,e_n &= \lambda_1\,e_n \end{aligned}$$

The above set of equations is written in the matrix form as shown below.

$$\begin{bmatrix} a_{11} & a_{12} & a_{13} & a_{14} & \ldots & a_{1n} \\ a_{21} & a_{22} & a_{23} & a_{24} & \ldots & a_{2n} \\ a_{31} & a_{32} & a_{33} & a_{34} & \ldots & a_{3n} \\ a_{41} & a_{42} & a_{43} & a_{44} & \ldots & a_{4n} \\ \ldots & \ldots & \ldots & \ldots & \ldots & \ldots \\ a_{n1} & a_{n2} & a_{n3} & a_{n4} & \ldots & a_{nn} \end{bmatrix} \begin{bmatrix} e_{11} & e_{21} & e_{31} & e_{41} & \ldots & e_{n1} \\ e_{12} & e_{22} & e_{23} & e_{42} & \ldots & e_{n2} \\ e_{13} & e_{23} & e_{33} & e_{43} & \ldots & e_{n3} \\ e_{14} & e_{24} & e_{34} & e_{44} & \ldots & e_{n4} \\ \ldots & \ldots & \ldots & \ldots & \ldots & \ldots \\ e_{1n} & e_{2n} & e_{3n} & e_{4n} & \ldots & e_{nn} \end{bmatrix}$$

$$= \begin{bmatrix} e_{11} & e_{21} & e_{31} & e_{41} & \ldots & e_{n1} \\ e_{12} & e_{22} & e_{23} & e_{42} & \ldots & e_{n2} \\ e_{13} & e_{23} & e_{33} & e_{43} & \ldots & e_{n3} \\ e_{14} & e_{24} & e_{34} & e_{44} & \ldots & e_{n4} \\ \ldots & \ldots & \ldots & \ldots & \ldots & \ldots \\ e_{1n} & e_{2n} & e_{3n} & e_{4n} & \ldots & e_{nn} \end{bmatrix} \begin{bmatrix} \lambda_1 & 0 & 0 & 0 & \ldots & 0 \\ 0 & \lambda_2 & 0 & 0 & \ldots & 0 \\ 0 & 0 & \lambda_3 & 0 & \ldots & 0 \\ 0 & 0 & 0 & \lambda_4 & \ldots & 0 \\ \ldots & \ldots & \ldots & \ldots & \ldots & \ldots \\ 0 & 0 & 0 & 0 & \ldots & \lambda_n \end{bmatrix}$$

$$\Rightarrow \begin{bmatrix} a_{11} & a_{12} & a_{13} & a_{14} & \ldots & a_{1n} \\ a_{21} & a_{22} & a_{23} & a_{24} & \ldots & a_{2n} \\ a_{31} & a_{32} & a_{33} & a_{34} & \ldots & a_{3n} \\ a_{41} & a_{42} & a_{43} & a_{44} & \ldots & a_{4n} \\ \ldots & \ldots & \ldots & \ldots & \ldots & \ldots \\ a_{n1} & a_{n2} & a_{n3} & a_{n4} & \ldots & a_{nn} \end{bmatrix}$$

$$= \begin{bmatrix} e_{11} & e_{21} & e_{31} & e_{41} & \dots & e_{n1} \\ e_{12} & e_{22} & e_{23} & e_{42} & \dots & e_{n2} \\ e_{13} & e_{23} & e_{33} & e_{43} & \dots & e_{n3} \\ e_{14} & e_{24} & e_{34} & e_{44} & \dots & e_{n4} \\ \dots & \dots & \dots & \dots & \dots & \dots \\ e_{1n} & e_{2n} & e_{3n} & e_{4n} & \dots & e_{nn} \end{bmatrix} \begin{bmatrix} \lambda_1 & 0 & 0 & 0 & \dots & 0 \\ 0 & \lambda_2 & 0 & 0 & \dots & 0 \\ 0 & 0 & \lambda_3 & 0 & \dots & 0 \\ 0 & 0 & 0 & \lambda_4 & \dots & 0 \\ \dots & \dots & \dots & \dots & \dots & \dots \\ 0 & 0 & 0 & 0 & \dots & \lambda_n \end{bmatrix} \times$$

$$\begin{bmatrix} e_{11} & e_{21} & e_{31} & e_{41} & \dots & e_{n1} \\ e_{12} & e_{22} & e_{23} & e_{42} & \dots & e_{n2} \\ e_{13} & e_{23} & e_{33} & e_{43} & \dots & e_{n3} \\ e_{14} & e_{24} & e_{34} & e_{44} & \dots & e_{n4} \\ \dots & \dots & \dots & \dots & \dots & \dots \\ e_{1n} & e_{2n} & e_{3n} & e_{4n} & \dots & e_{nn} \end{bmatrix}^T$$

$$\text{Where A} = \begin{matrix} a_{11} & a_{12} & a_{13} & a_{14} & \dots & a_{1n} \\ a_{21} & a_{22} & a_{23} & a_{24} & \dots & a_{2n} \\ a_{31} & a_{32} & a_{33} & a_{34} & \dots & a_{3n} \\ a_{41} & a_{42} & a_{43} & a_{44} & \dots & a_{4n} \\ \dots & \dots & \dots & \dots & \dots & \dots \\ a_{n1} & a_{n2} & a_{n3} & a_{n4} & \dots & a_{nn} \end{matrix}$$

$$\text{Eigen vectors is given as } e_1 = \begin{bmatrix} e_{11} \\ e_{12} \\ e_{13} \\ e_{14} \\ e_{15} \\ \dots \\ e_{1n} \end{bmatrix} e_2 = \begin{bmatrix} e_{21} \\ e_{32} \\ e_{23} \\ e_{24} \\ e_{25} \\ \dots \\ e_{2n} \end{bmatrix} e_3 = \begin{bmatrix} e_{31} \\ e_{32} \\ e_{33} \\ e_{34} \\ e_{35} \\ \dots \\ e_{3n} \end{bmatrix} \dots e_n = \begin{bmatrix} e_{n1} \\ e_{n2} \\ e_{n3} \\ e_{n4} \\ e_{n5} \\ \dots \\ e_{nn} \end{bmatrix}$$

$$\text{The Eigen vector matrix E} = \begin{matrix} e_{11} & e_{21} & e_{31} & e_{41} & \dots & e_{n1} \\ e_{12} & e_{22} & e_{23} & e_{42} & \dots & e_{n2} \\ e_{13} & e_{23} & e_{33} & e_{43} & \dots & e_{n3} \\ e_{14} & e_{24} & e_{34} & e_{44} & \dots & e_{n4} \\ \dots & \dots & \dots & \dots & \dots & \dots \\ e_{1n} & e_{2n} & e_{3n} & e_{4n} & \dots & e_{nn} \end{matrix}$$

$$\text{Diagonal matrix D} = \begin{bmatrix} \lambda_1 & 0 & 0 & 0 & \dots & 0 \\ 0 & \lambda_2 & 0 & 0 & \dots & 0 \\ 0 & 0 & \lambda_3 & 0 & \dots & 0 \\ 0 & 0 & 0 & \lambda_4 & \dots & 0 \\ \dots & \dots & \dots & \dots & \dots & \dots \\ 0 & 0 & 0 & 0 & \dots & \lambda_n \end{bmatrix}$$

Thus the non-deficient matrix A is represented as the product of the transpose of the Eigen vector matrix, Diagonal matrix and the Eigen vector matrix (i.e.) $A = EDE^T$.

1.21 Schur's Lemma

For any square matrix A, there exists the Unitary matrix U such that $U^H AU = T$, where T is the triangular matrix.

Example 1.21. Consider the matrix A as shown below.

$$A = \begin{bmatrix} 1 & 2 & 3 \\ 4 & 5 & 7 \\ 7 & 8 & 10 \end{bmatrix}$$

Eigen values and the corresponding Eigen vectors are as shown below.

17.1747, −1, −0.1747. The Eigen vector corresponding to the Eigen value 17.1747 is $E_1 = \begin{bmatrix} -0.2176 \\ -0.5392 \\ -0.8136 \end{bmatrix}$.

Construct the Unitary matrix U_1 with the Eigen vector E_1 as the first column and other two columns are arbitrarily chosen.

$$U_1 = \begin{bmatrix} -0.2176 & 1 & -1 \\ -0.5392 & 1 & 0.7726 \\ -0.8136 & -0.9302 & -0.2445 \end{bmatrix}$$

$$(U_1)^H A\, U_1 = \begin{bmatrix} -0.2176 & 1 & -1 \\ -0.5392 & 1 & 0.7726 \\ -0.8136 & -0.9302 & -0.2445 \end{bmatrix}^H \times$$

$$\begin{bmatrix} 1 & 2 & 3 \\ 4 & 5 & 7 \\ 7 & 8 & 10 \end{bmatrix} \begin{bmatrix} -0.2176 & 1 & -1 \\ -0.5392 & 1 & 0.7726 \\ -0.8136 & -0.9302 & -0.2445 \end{bmatrix}$$

$$= \begin{bmatrix} 17.1753 & -6.0233 & 3.6934 \\ 0 & -2.6023 & 0.9996 \\ 0 & 0.3201 & -0.4418 \end{bmatrix}$$

Now consider the matrix $B = \begin{bmatrix} -2.6023 & 0.9996 \\ 0.3201 & -0.4418 \end{bmatrix}$

The Eigen values of the matrix B are −2.7414 and −0.3027. The corresponding Eigen vector −2.7414 is $\begin{bmatrix} -0.9905 \\ 0.1379 \end{bmatrix}$

Form the matrix $U_2 = \begin{bmatrix} 1 & 0 & 0 \\ 0 & -0.9905 & e1 \\ 0 & 0.1379 & e2 \end{bmatrix}$

The vector $\begin{bmatrix} e1 \\ e2 \end{bmatrix}$ are chosen such that the columns of the matrix $\begin{bmatrix} -0.9905 & e1 \\ 0.1379 & e2 \end{bmatrix}$ are orthogonal to each other.

Thus the matrix U_2 is chosen as $\begin{bmatrix} 1 & 0 & 0 \\ 0 & -0.9905 & 1 \\ 0 & 0.1379 & 7.1827 \end{bmatrix}$

$$U_2'^{*} U_2 = \begin{bmatrix} 1 & 0 & 0 \\ 0 & 1 & 0 \\ 0 & 0 & 52.5912 \end{bmatrix}$$

$$(U_2)^H (U_1)^H A\, U_1 U_2 = \begin{bmatrix} 17.1753 & 6.4754 & 20.5055 \\ 0 & -2.7417 & -4.9272 \\ 0 & 0 & -15.9140 \end{bmatrix}$$

Let $U^H = (U_2)^H (U_1)^H$, *which is also the Unitary Matrix*

$$\Rightarrow U^H A\, U = \begin{bmatrix} 17.1753 & 6.4754 & 20.5055 \\ 0 & -2.7417 & -4.9272 \\ 0 & 0 & -15.9140 \end{bmatrix}$$

By using the procedure described above, any arbitrary matrix A, there exists the Unitary matrix U such that $U^H AU = T$, where T is the triangular matrix.

1.22 Hermitian Matrices and Skew Hermitian Matrices

Let A^H is the conjugate transpose of the matrix A. The matrix is said to be Hermitian matrix if $A^H = A$ and the matrix A is said to be Skew Hermitian if $A^H = -A$.

Properties:

1. $x^H Ax$ is real.

   Proof. Taking the conjugate transpose of the matrix $x^H Ax$, we get $(x^H Ax)^H = x^H Ax$. Hence proved.

2. Eigen values of the Hermitian matrix are real.

Proof. Vector such that $Ax = \lambda x$.
Multiplying x^H on both sides, we get

$$x^H Ax = \lambda x^H x.$$

We know $x^H Ax$ and $x^H x$ are the real numbers.

$$\Rightarrow \lambda\, x^H x = real\ number$$
$$\Rightarrow \lambda \text{ is the real number.}$$

3. Eigen vectors of the Hermitian matrix A corresponding to distinct Eigen values are orthogonal.

Proof. Let A be the Hermitian matrix and let λ_1, λ_2 be the distinct Eigen values of the matrix A and the corresponding Eigen vectors are e_1 and e_2.

$$Ae_1 = \lambda_1 e_1$$
$$Ae_2 = \lambda_2 e_2$$

Consider

$$(\lambda_1 e_1)^H e_2 = (A\, e_1)^H e_2 = e_1{}^H A^H e_2 = e_1{}^H Ae_2 = e_1{}^H \lambda_2 e_2$$

(Note that $A = A^H$)

$$\Rightarrow (\lambda_1 e_1)^H e_2 = e_1{}^H \lambda_2 e_2 \Rightarrow e_1{}^H e_2(\lambda_1 - \lambda_2) = 0$$
$$\Rightarrow e_1{}^H e_2 = 0\ [Because\ \lambda_1 - \lambda_2 \neq 0]$$

Hence proved.

Example 1.22. Let A = $\begin{bmatrix} 1 & i & 3i \\ -i & 2 & 4 \\ -3i & 4 & 5 \end{bmatrix}$ be the Hermitian matrix (i.e.) A = A^H

Eigen values are $-1.3560, 0.4123$ and 8.9438 (They are real) and the corresponding Column wise Eigen vectors are

$$\begin{bmatrix} 0.5410i & -0.7601i & 0.3599i \\ 0.5728 & 0.6464 & 0.5041 \\ -0.6158 & -0.0666 & 0.7851 \end{bmatrix}$$

Note that Eigen vectors are orthogonal to each other. (i.e.)

$$\begin{bmatrix}-0.5410i & 0.5728 & -0.6158\end{bmatrix}\begin{bmatrix}0.5410i \\ 0.5728 \\ -0.6158\end{bmatrix} = -1.1102^{-16} \simeq 0$$

Similarly it can be shown that all the Eigen vectors are orthogonal to each other.

4. Hermiitian matrix is always diagonalizable using the unitary matrix.
For all arbitrary Hermitian matrixes 'A', there exists the Unitary matrix U such $U^H AU = D$, where D is the Diagonal matrix.

Proof. From Schur's lemma, for any arbitrary Hermitian matrix 'A', there exists the Unitary matrix U such that $U^H AU = T$. Taking Hermitian transpose on both sides, we get

$$U^H A^H U = T^H$$
$$\Rightarrow U^H AU = T^H \quad [\text{Because A} = A^H]$$
$$\Rightarrow T = T^H$$

⇒ 'T', in this case is the Diagonal matrix 'D' with all the elements in the matrix are filled up with real numbers. Hence proved.

1.23 Unitary Matrices

Columns of the unitary matrices are orthonormal to each other. Let U be the Unitary matrix, then $U^H U = I$ (Identity matrix)

Properties:

1. $\|Ux\| = \|x\|$.

Proof.

$$\text{Sqrt}((Ux)^H (Ux)) = Sqrt((x)^H (U)^H (U)(x)) = Sqrt((x)^H (x)) = \|x\|$$

Hence proved.

2. $(Ux)^H ((Uy) = (x)^H (y)$.
3. If U_1, U_2 are Unitary matrices, then $U_1 U_2$ is also the Unitary matrix.
4. U is always invertible. In particular the inverse of the unitary matrix U is given as U^H.
5. Magnitude of the Eigen values of the unitary matrix is always one.

Proof. The Eigen vector x of the unitary matrix 'U' satisfies the condition $Ux = \lambda x$, where λ is the Eigen value of the unitary matrix

Taking norm on both sides, we get

$$
\begin{aligned}
&\|Ux\| = \|x\| \text{ (see property 1)}\\
&\|Ux\| = \|\lambda x\| = sqrt((\lambda x)^H(\lambda x))\\
&= sqrt(x^H\lambda^H\lambda x) = sqrt((\lambda^H\lambda)sqrt((x^H x)) = \|\lambda\|\|x\|\\
&= |\lambda|\|x\|\\
&\Rightarrow \|x\| = |\lambda|\|x\|\\
&\Rightarrow |\lambda| = 1
\end{aligned}
$$

Hence proved.

6. Eigen vectors corresponding to distinct Eigen values are orthogonal.

Proof. Consider the Unitary matrix U such that $Ux_1 = \lambda_1 x_1$ and $x_2 = \lambda_2 x_2$, where x_1, x_2 are Eigen vectors corresponding to the distinct Eigen values λ_1 and λ_2.

Consider $(\lambda_1 x_1)^H(\lambda_2 x_1) = \overline{\lambda_1}\lambda_2 {x_1}^H x_1 = (Ux_1)^H(Ux_1) = {x_1}^H x_1$

$$
\begin{aligned}
&\Rightarrow \overline{\lambda_1}\lambda_2 {x_1}^H x_1 = {x_1}^H x_1\\
&\Rightarrow (\overline{\lambda_1}\lambda_2 - 1){x_1}^H x_1 = 0\\
&\Rightarrow {x_1}^H x_1 = 0 \text{ [Because } (\overline{\lambda_1}\lambda_2 - 1) \neq 0]
\end{aligned}
$$

Hence proved

Example 1.23.

$$A = \begin{bmatrix} 0.7071 & 0.7071 \\ 0.7071i & -0.7071i \end{bmatrix}$$

1. Note that the columns of the matrix are orthonormal to each other.

$$\text{(i.e.) } A^H A = \begin{bmatrix} 1 & 0 \\ 0 & 1 \end{bmatrix}$$

2. The Eigen values of the matrix A are 0.9659 + 0.2588i and 0.2588 − 0.9659i. Note that the magnitude of the Eigen values are 1.
3. The Eigen vectors corresponding to the distinct Eigen values are listed below.

$$E_1 = \begin{vmatrix} 0.8881 \\ 0.3251 + 0.3251i \end{vmatrix}, \; E_2 = \begin{bmatrix} -0.3251 + 0.3251i \\ 0.8881 \end{bmatrix}$$

Note that they are orthonormal to each other.

$$\text{(i.e) } {E_1}^H E_2 = \begin{bmatrix} 1 & 0 \\ 0 & 1 \end{bmatrix}$$

Example 1.24. The DFT matrix is the unitary matrix.

Four-point DFT matrix is as shown below.

$$A = \left(\frac{1}{2}\right)\begin{bmatrix} 1 & 1 & 1 & 1 \\ 1 & w & w^2 & w^3 \\ 1 & w^2 & w^4 & w^6 \\ 1 & w^3 & w^6 & w^9 \end{bmatrix}$$

where $w = e^{\left(\frac{j2\pi}{4}\right)}$

Note that the columns of the matrix A are orthonormal to each other.

$$(i.e)A' * A = \begin{bmatrix} 1 & 0 & 0 & 0 \\ 0 & 1 & 0 & 0 \\ 0 & 0 & 1 & 0 \\ 0 & 0 & 0 & 1 \end{bmatrix}$$

1. The Eigen values of the matrix A are 1, −1, i
2. The Eigen vectors corresponding to the above mentioned Eigen values are given as

$$\begin{bmatrix} 0.8660 \\ 0.2887 \\ 0.2887 \\ 0.2887 \end{bmatrix}, \begin{bmatrix} -0.5 \\ 0.5 \\ 0.5 \\ 0.5 \end{bmatrix}, \begin{bmatrix} 0 \\ 0.7071 \\ 0 \\ -0.7071 \end{bmatrix}, \begin{bmatrix} -0.0231 - 0.0070i \\ -0.4158 - 0.0035i \\ 0.8085 \\ -0.4158 - 0.0035i \end{bmatrix}$$

Characteristics of the DFT Matrices

1. The Eigen values of the DFT matrices are one among the following values 1, −1, i, −i. The magnitude of the Eigen values are always one.
2. Consider the circular convolution of the following two sequences as shown below.

$$\text{Let X} = \begin{bmatrix} 1 \\ 2 \\ 3 \\ 4 \end{bmatrix} \text{ and H} = \begin{bmatrix} 5 \\ 6 \\ 7 \\ 8 \end{bmatrix}$$

The circular convolution performed using matrix method is as shown below.

$$\begin{bmatrix} 1 & 4 & 3 & 2 \\ 2 & 1 & 4 & 3 \\ 3 & 2 & 1 & 4 \\ 4 & 3 & 2 & 1 \end{bmatrix}\begin{bmatrix} 5 \\ 6 \\ 7 \\ 8 \end{bmatrix} = \begin{bmatrix} 66 \\ 68 \\ 66 \\ 60 \end{bmatrix}$$

The circular matrix $C = \begin{bmatrix} 1 & 4 & 3 & 2 \\ 2 & 1 & 4 & 3 \\ 3 & 2 & 1 & 4 \\ 4 & 3 & 2 & 1 \end{bmatrix}$ is diagonalized using DFT Unitary matrix U as shown below.

$$C = UDU^{-1}$$

$$\Rightarrow \begin{bmatrix} 1 & 4 & 3 & 2 \\ 2 & 1 & 4 & 3 \\ 3 & 2 & 1 & 4 \\ 4 & 3 & 2 & 1 \end{bmatrix} = \begin{bmatrix} 0.5 & 0.5 & 0.5 & 0.5 \\ 0.5 & 0.5i & -0.5 & -0.5i \\ 0.5 & -0.5 & 0.5 & -0.5 \\ 05 & -0.5i & -0.5 & 0.5i \end{bmatrix} \times$$

$$\begin{bmatrix} 10 & 0 & 0 & 0 \\ 0 & -2+2i & 0 & 0 \\ 0 & 0 & -2-2i & 0 \\ 0 & 0 & 0 & -2 \end{bmatrix} \begin{bmatrix} 0.5 & 0.5 & 0.5 & 0.5 \\ 0.5 & -0.5i & -0.5 & 0.5i \\ 0.5 & -0.5 & 0.5 & -0.5 \\ 0.5 & 0.5i & -0.5 & -0.5i \end{bmatrix}$$

Where $U = \begin{bmatrix} 0.5 & 0.5 & 0.5 & 0.5 \\ 0.5 & 0.5i & -0.5 & -0.5i \\ 0.5 & -0.5 & 0.5 & -0.5 \\ 05 & -0.5i & -0.5 & 0.5i \end{bmatrix}$

$$D = \begin{bmatrix} 10 & 0 & 0 & 0 \\ 0 & -2+2i & 0 & 0 \\ 0 & 0 & -2-2i & 0 \\ 0 & 0 & 0 & -2 \end{bmatrix}$$

Let $B = \begin{bmatrix} 5 \\ 6 \\ 7 \\ 8 \end{bmatrix}$

Multiplying Matrix B on both sides in the above equation we get,

$$CB = UDU^{-1}B$$

$$\Rightarrow \begin{bmatrix} 1 & 4 & 3 & 2 \\ 2 & 1 & 4 & 3 \\ 3 & 2 & 1 & 4 \\ 4 & 3 & 2 & 1 \end{bmatrix} \begin{bmatrix} 5 \\ 6 \\ 7 \\ 8 \end{bmatrix} = \begin{bmatrix} 0.5 & 0.5 & 0.5 & 0.5 \\ 0.5 & 0.5i & -0.5 & -0.5i \\ 0.5 & -0.5 & 0.5 & -0.5 \\ 05 & -0.5i & -0.5 & 0.5i \end{bmatrix} \times$$

$$\begin{bmatrix} 10 & 0 & 0 & 0 \\ 0 & -2+2i & 0 & 0 \\ 0 & 0 & -2-2i & 0 \\ 0 & 0 & 0 & -2 \end{bmatrix} \begin{bmatrix} 0.5 & 0.5 & 0.5 & 0.5 \\ 0.5 & -0.5i & -0.5 & 0.5i \\ 0.5 & -0.5 & 0.5 & -0.5 \\ 0.5 & 0.5i & -0.5 & -0.5i \end{bmatrix} \begin{bmatrix} 5 \\ 6 \\ 7 \\ 8 \end{bmatrix}$$

The product $U^{-1}B$ in the RHS as shown below is given as

$$\begin{bmatrix} 0.5 & 0.5 & 0.5 & 0.5 \\ 0.5 & -0.5i & -0.5 & 0.5i \\ 0.5 & -0.5 & 0.5 & -0.5 \\ 0.5 & 0.5i & -0.5 & -0.5i \end{bmatrix} \begin{bmatrix} 5 \\ 6 \\ 7 \\ 8 \end{bmatrix} = \begin{bmatrix} 13 \\ -1+i \\ -1 \\ -1-i \end{bmatrix}$$

. This can be viewed as the DFT of the vector 'B' (Except the scaling factor). The vector thus obtained is in frequency domain. The vector thus obtained in frequency domain is scaled using the diagonal matrix D as shown below. $DU^{-1}B$

$$\Rightarrow \begin{bmatrix} 10 & 0 & 0 & 0 \\ 0 & -2+2i & 0 & 0 \\ 0 & 0 & -2-2i & 0 \\ 0 & 0 & 0 & -2 \end{bmatrix} \begin{bmatrix} 13 \\ -1+i \\ -1 \\ -1-i \end{bmatrix} = \begin{bmatrix} 1.3X10^2 \\ -0.040iX10^2 \\ 0.02X10^2 \\ 0.04iX10^2 \end{bmatrix}$$

The obtained vector is then multiplied with the Unitary matrix U can be viewed as the inverse DFT as shown below to obtain the circular convolved output.

$$\begin{bmatrix} 0.5 & 0.5 & 0.5 & 0.5 \\ 0.5 & 0.5i & -0.5 & -0.5i \\ 0.5 & -0.5 & 0.5 & -0.5 \\ 05 & -0.5i & -0.5 & 0.5i \end{bmatrix} \begin{bmatrix} 1.3X10^2 \\ -0.040iX10^2 \\ 0.02X10^2 \\ 0.04iX10^2 \end{bmatrix} = \begin{bmatrix} 66 \\ 68 \\ 66 \\ 60 \end{bmatrix}$$

1.24 Normal Matrices

The matrix A that satisfies the condition $A^H A = AA^H$ is called Normal Matrix. Hermitian matrix, Unitary matrix, Permutation matrix and circular matrices are called Normal matrices. All the Normal matrices are diagonalizable.

Summary (see Fig. 1.6 below)

1. For any matrix A of size m × m, the Eigen vectors corresponding to distinct Eigen values are independent. The Eigen vectors corresponding to the particular Eigen value are independent, because it is the basis for the Null space of the matrix of the form $\|[A - \lambda I]$. So in case of non-deficient matrix A, the Eigen vectors forms the basis for the vector space $\mathbb{R}^m$.
2. All the Normal matrices are non-defective. (i.e.) Any Normal matrix A of size m × m can be represented as the product of UDU^H, where U is the unitary matrix and D is the Diagonal matrix. The Eigen vectors corresponding to various Eigen values are orthogonal to each other. Hermitian Matrix, Unitary Matrix (Example DFT Matrix), Permutation matrix and the Circular matrix are the examples for the Normal matrices. All Normal matrices need not be invertible.

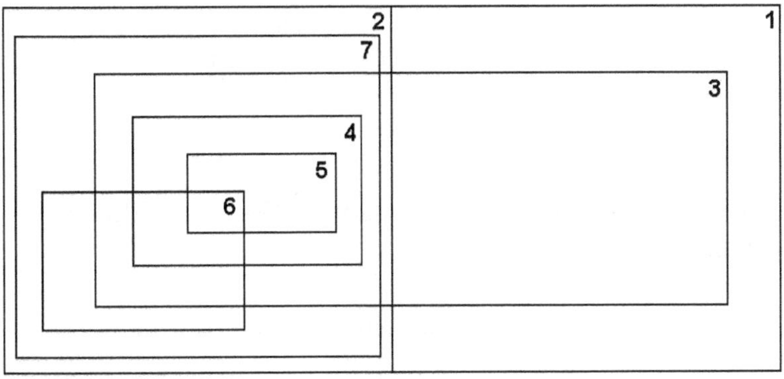

Fig. 1.6 Set of deficient and non-deficient matrices
1. Deficient Matrices
2. Non-Deficient Matrices
3. Invertible Matrices
4. Unitary Matrices
5. Permutation Matrices
6. Hermitian Matrices
7. Normal Matrices

3. The Eigen values of the Hermitian matrix A of size m × m are real. Any arbitrary Hermitian matrix can be represented as the product of UDU^H, where U is the unitary matrix and D is the Diagonal matrix. The Eigen vectors of the Hermitian matrix 'A' forms the orthogonal basis of the vector space $\mathbb{R}^m$. Hermitian matrix need not be invertible.
4. The magnitude of the Eigen values of the Unitary matrix A of size m × m is 1. The value can be complex. This matrix can also be represented always as UDU^H. where U is the unitary matrix and D is the Diagonal matrix. The Eigen vectors of the Unitary matrix 'A' forms the orthogonal basis of the vector space $\mathbb{R}^m$. Unitary matrix is always invertible.
5. DFT matrix is the example for the Unitary matrix which is the type Normal matrix. DFT matrix is always diagonalizable (i.e.) non-deficient. The DFT matrix A of size m × m can be represented as UDU^H, where U is the unitary matrix and D is the Diagonal matrix. The Eigen values of the matrix A hold any of the following values i, −I, 1, −1. DFT matrix is the Unitary matrix and is always invertible. The Eigen vectors of the DFT matrix 'A' forms the orthogonal basis of the vector space $\mathbb{R}^m$.
6. Permutation matrix is the example for the unitary matrix, which is the type of the Normal matrix and hence diagonalizable (i.e.) non-deficient. The permutation matrix A of size m × m can be represented as UDU^H. The magnitude of the Eigen values of the permutation matrix is 1. Permutation matrix, which is the unitary matrix, is always invertible. The Eigen vectors of the Permutation matrix 'A' of size m × m forms the orthogonal basis of the vector space $\mathbb{R}^m$.
7. Circular matrix is the type of Normal matrix and hence diagonalizable (i.e.) non-deficient. The circular matrix A of size m × m can be represented as U^H, where

U is the unitary matrix and D is the Diagonal matrix. Circular matrix need not be invertible matrix. The Eigen vectors of the Circular matrix 'A' forms the orthogonal basis of the vector space $\mathbb{R}^m$.

1.25 Applications of Diagonalization of the Non-deficient Matrix

(a) Solving the Difference Equation of the form $U_{k+1} = AU_k$, where

$$U_{k+1} = \begin{bmatrix} x1(k+1) \\ x2(k+2) \\ x3(k+3) \\ \cdots \\ xn(k+1) \end{bmatrix} \quad U_k = \begin{bmatrix} x1(k) \\ x2(k) \\ x3(k) \\ \cdots \\ xn(k) \end{bmatrix}$$

Where 'A' is the non-deficient n × n matrix.

Example 1.25. Let the non-deficient 2 × 2 matrix be A $= \begin{bmatrix} 1 & 1 \\ 1 & 0 \end{bmatrix}$.

$$U_{k+1} = \begin{bmatrix} x1(k+1) \\ x2(k+2) \end{bmatrix} \; and \; U_k = \begin{bmatrix} x1(k) \\ x2(k) \end{bmatrix}$$

Solving $U_{k+1} = AU_k$ is equivalent to solving the equation $U_k = A^k U_0$.
Note that the matrix A is the unitary matrix. Therefore the matrix A can be represented as $A = ED\,E^H$ (as given below), where E is the Eigen matrix, in which the columns are the Eigen vectors.

$$\begin{bmatrix} 1 & 1 \\ 1 & 0 \end{bmatrix}$$

$$= \begin{bmatrix} 0.5257 & -0.8507 \\ -0.8507 & -0.5257 \end{bmatrix} \begin{bmatrix} -0.6180 & 0 \\ 0 & 1.6180 \end{bmatrix} \begin{bmatrix} 0.5257 & -0.8507 \\ -0.8507 & -0.5257 \end{bmatrix}$$

$$\Rightarrow [A]^k = EDE^H EDE^H EDE^H EDE^H \ldots (\text{'}k\text{' } times)$$

Note that EE^H is the Identity matrix and hence

$$[A]^k = EDDD \ldots (k \; times) E^H$$

$$\Rightarrow [A]^k = ED^k E^H$$

$$Also \; D^k = \begin{bmatrix} (-0.6180)^k & 0 \\ 0 & (1.6180)^k \end{bmatrix}$$

Thus solution to the above difference equation is given as

$$U_k = A^k U_0 = ED^k E^H U_0$$
$$= \begin{bmatrix} 0.5257 & -0.8507 \\ -0.8507 & -0.5257 \end{bmatrix} \begin{bmatrix} (-0.6180)^k & 0 \\ 0 & (1.6180)^k \end{bmatrix} \begin{bmatrix} 0.5257 & -0.8507 \\ -0.8507 & -0.5257 \end{bmatrix} U_0$$

If $U_0 = \begin{bmatrix} 1 \\ 0 \end{bmatrix}$

$$U_k = \begin{bmatrix} 0.5257 & -0.8507 \\ -0.8507 & -0.5257 \end{bmatrix} \begin{bmatrix} (-0.6180)^k & 0 \\ 0 & (1.6180)^k \end{bmatrix} \begin{bmatrix} 0.5257 & -0.8507 \\ -0.8507 & -0.5257 \end{bmatrix} \begin{bmatrix} 1 \\ 0 \end{bmatrix}$$
$$= \begin{bmatrix} 0.5257 & -0.8507 \\ -0.8507 & -0.5257 \end{bmatrix} \begin{bmatrix} (-0.6180)^k & 0 \\ 0 & (1.6180)^k \end{bmatrix} \begin{bmatrix} 0.5257 \\ -0.8507 \end{bmatrix}$$
$$= \begin{bmatrix} 0.5257 & -0.8507 \\ -0.8507 & -0.5257 \end{bmatrix} \begin{bmatrix} (0.5257)(-0.6180)^k \\ (-0.8507)(1.6180)^k \end{bmatrix}$$
$$= \begin{bmatrix} 0.5257 & -0.8507 \\ -0.8507 & -0.5257 \end{bmatrix} \begin{bmatrix} (0.5257)(-0.6180)^k \\ (-0.8507)(1.6180)^k \end{bmatrix}$$
$$= \begin{bmatrix} [(0.5257)^2(-0.6180)^k + (-0.8507)^2(1.6180)^k] \\ (0.5257)(-0.8507)(-0.6180)^k + (0.5257)(-0.8507)(1.6180)^k \end{bmatrix}$$
$$= \begin{bmatrix} [(0.2764)(-0.6180)^k + (0.7237)(1.6180)^k] \\ [(-0.4472)(-0.6180)^k + (0.4472)(1.6180)^k \end{bmatrix}$$

(b) Solving Differential Equation of the form

$\frac{dU(t)}{dt} = \text{A U(t)}$, where $U(t) = \begin{bmatrix} u_1(t) \\ u_2(t) \\ \cdots \\ u_n(t) \end{bmatrix}$ and A is the Non-deficient matrix.

Solution for the above differential equation is of the following form

$$\text{U(t)} = e^{At}\text{U(0)}$$

Let us use the non-deficient matrix $A = \begin{bmatrix} 1 & 1 \\ 1 & 0 \end{bmatrix}$ and $\text{U(0)} = \begin{bmatrix} 1 \\ 0 \end{bmatrix}$

To compute e^{At}, Diagonalization of the matrix 'A' is used.

$$e^A = I + \frac{A}{1!} + \frac{(A)^2}{2!} + \frac{(A)^3}{3!} + \cdots$$
$$\Rightarrow e^A = I + \frac{A}{1!} + \frac{EDE^H EDE^H}{2!} + \frac{EDE^H EDE^H EDE^H}{3!} + \cdots$$
$$\Rightarrow e^A = EIE^H + \frac{EDE^H}{1!} + \frac{EDDE^H}{2!} + \frac{EDDDE^H}{3!} + \cdots$$

$$\Rightarrow e^{A} = E\left(I + \frac{D}{1!} + \frac{DD}{2!} + \frac{DDD}{3!} + \cdots\right)E^{H}$$
$$\Rightarrow e^{A} = Ee^{D}E^{H}$$
$$\Rightarrow e^{At} = Ee^{Dt}E^{H}$$

For the given problem the solution of the differential equation is given as follows.

$$\mathrm{U(t)} = e^{At}\mathrm{U(0)}$$
$$\Rightarrow U(t) = Ee^{Dt}E^{H}U(0)$$
$$\Rightarrow U(t) = \begin{bmatrix} 0.5257 & -0.8507 \\ -0.8507 & -0.5257 \end{bmatrix} e^{Dt} \begin{bmatrix} 0.5257 & -0.8507 \\ -0.8507 & -0.5257 \end{bmatrix}^{H} \begin{bmatrix} 1 \\ 0 \end{bmatrix}$$

where $\mathrm{D} = \begin{bmatrix} -0.6180 & 0 \\ 0 & 1.6180 \end{bmatrix}$

$$\Rightarrow e^{Dt} = e^{\begin{bmatrix} -0.6180 & 0 \\ 0 & 1.6180 \end{bmatrix} t} = \begin{bmatrix} e^{-0.6180t} & 0 \\ 0 & e^{1.6180t} \end{bmatrix}$$

$$\Rightarrow U(t)$$
$$= \begin{bmatrix} 0.5257 & -0.8507 \\ -0.8507 & -0.5257 \end{bmatrix} \begin{bmatrix} e^{-0.6180t} & 0 \\ 0 & e^{1.6180t} \end{bmatrix} \begin{bmatrix} 0.5257 & -0.8507 \\ -0.8507 & -0.5257 \end{bmatrix} \begin{bmatrix} 1 \\ 0 \end{bmatrix}$$
$$= \begin{bmatrix} 0.5257 & -0.8507 \\ -0.8507 & -0.5257 \end{bmatrix} \begin{bmatrix} 0.5257e^{-0.6180t} \\ (-0.8507)e^{1.6180t} \end{bmatrix}$$
$$= \begin{bmatrix} 0.5257 & -0.8507 \\ -0.8507 & -0.5257 \end{bmatrix} \begin{bmatrix} 0.5257e^{-0.6180t} \\ (-0.8507)e^{1.6180t} \end{bmatrix}$$
$$= \begin{bmatrix} 0.5257 & -0.8507 \\ -0.8507 & -0.5257 \end{bmatrix} \begin{bmatrix} 0.5257e^{-0.6180t} \\ (-0.8507)e^{1.6180t} \end{bmatrix}$$
$$= \begin{bmatrix} 0.2764e^{-0.6180t} + 0.7234e^{1.6180t} \\ -0.4472e^{-0.6180t} + 0.4472e^{1.6180t} \end{bmatrix}$$

1.26 Singular Value Decomposition

Consider the matrix A of size m × n. The matrix A can be represented as the product of Hermitian transpose of the unitary matrix U_1, Diagonal matrix 'D' and the unitary matrix 'U_2'. $A = U_1 \mathrm{D} {U_2}^{H}$.

Let the unit magnitude Eigen vector 'v_i' corresponding to the matrix $A^T A$ satisfies the condition $A^T A\ v_i = \lambda_i\ v_i$. Multiplying A on both sides, we get $AA^T(A\ v_i) = \lambda_i(A\ v_i)$. $\Rightarrow$ $(A\ v_i)$ is the Eigen vector of the matrix AA^T. The corresponding Eigen value is λ_i. The magnitude of the vector Av_i is obtained as $(Av_i)^T(Av_i) = {v_i}^T A^T Av_i = \lambda_i {v_i}^T v_i = \lambda_i$. The Unit magnitude Eigen vector of the matrix AA^T can be represented as $Av_i/\sqrt{\lambda_i}$. Let it be u_i

Rewriting the above equation as $Av_i = \sqrt{\lambda_i}\, u_i$

Example 1.26. Consider the matrix A $= \begin{bmatrix} 1 & 2 & 3 \\ 4 & 5 & 6 \end{bmatrix}$.

$$AA^T = \begin{bmatrix} 14 & 32 \\ 32 & 77 \end{bmatrix}$$

The Eigen values of the matrix AA^T are 0.5973, 90.4027 and the corresponding Eigen vectors are given as $\begin{bmatrix} -0.9224 \\ 0.3863 \end{bmatrix}, \begin{bmatrix} 0.3863 \\ 0.9224 \end{bmatrix}$

Similarly the Eigen values of the matrix $A^T A$ are 0, 0.5963, 90.4027 and the corresponding Eigen vectors are given as

$$\begin{bmatrix} -0.4082 \\ 0.8165 \\ -0.4082 \end{bmatrix}, \begin{bmatrix} -0.8060 \\ -0.1124 \\ 0.5812 \end{bmatrix} \text{ and } \begin{bmatrix} 0.4287 \\ 0.5663 \\ 0.7039 \end{bmatrix}.$$

The Eigen vectors of the matrix AA^T are obtained as following for the corresponding non-zero Eigen values.

$$\frac{\begin{bmatrix} 1 & 2 & 3 \\ 4 & 5 & 6 \end{bmatrix} \begin{bmatrix} -0.8060 \\ -0.1124 \\ 0.5812 \end{bmatrix}}{\sqrt{0.5963}} = \begin{bmatrix} 0.9224 \\ -0.3863 \end{bmatrix}$$

$$\frac{\begin{bmatrix} 1 & 2 & 3 \\ 4 & 5 & 6 \end{bmatrix} \begin{bmatrix} -0.4287 \\ 0.5663 \\ 0.7039 \end{bmatrix}}{\sqrt{90.4027}} = \begin{bmatrix} 0.3863 \\ 0.9224 \end{bmatrix}$$

Note that it is same as the one computed directly from the basic definition. Also note that the Eigen vectors thus obtained are orthonormal to each other as the matrix AA^T and $A^T A$ are Hermitian matrix.

Thus the matrix

$$\begin{bmatrix} 1 & 2 & 3 \\ 4 & 5 & 6 \end{bmatrix} \begin{bmatrix} -0.8060 & -0.4287 \\ 0.1124 & 0.5663 \\ 0.5812 & 0.7039 \end{bmatrix}$$

$$= \begin{bmatrix} 0.9224 & 0.3863 \\ -0.3863 & 0.9224 \end{bmatrix} \begin{bmatrix} \sqrt{0.5963} & 0 \\ 0 & \sqrt{90.4027} \end{bmatrix}$$

As the Eigen vectors are orthonormal, the matrix A can be represented as follows as follows

$$\Rightarrow \begin{bmatrix} 1 & 2 & 3 \\ 4 & 5 & 6 \end{bmatrix}$$

$$= \begin{bmatrix} 0.9224 & 0.3863 \\ -0.3863 & 0.9224 \end{bmatrix} \begin{bmatrix} \sqrt{0.5963} & 0 \\ 0 & \sqrt{90.4027} \end{bmatrix} \begin{bmatrix} -0.8060 & -0.4287 \\ -0.1124 & 0.5663 \\ 0.5812 & 0.7039 \end{bmatrix}^T$$

$$= \begin{bmatrix} 0.9224 & 0.3863 \\ -0.3863 & 0.9224 \end{bmatrix} \begin{bmatrix} \sqrt{0.5963} & 0 \\ 0 & \sqrt{90.4027} \end{bmatrix} \begin{bmatrix} -0.8060 & -0.4287 \\ -0.1124 & 0.5663 \\ 0.5812 & 0.7039 \end{bmatrix}^T$$

We can even include the Eigen vector corresponding to the zero Eigen value of the matrix $A^T A$ as shown below.

$$\Rightarrow \begin{bmatrix} 1 & 2 & 3 \\ 4 & 5 & 6 \end{bmatrix}$$

$$= \begin{bmatrix} 0.9224 & 0.3863 \\ -0.3863 & 0.9224 \end{bmatrix} \begin{bmatrix} \sqrt{0.5963} & 0 & 0 \\ 0 & \sqrt{90.4027} & 0 \\ 0 & 0 & 0 \end{bmatrix} \begin{bmatrix} -0.8060 & -0.4287 & -0.4082 \\ -0.1124 & 0.5663 & 0.8165 \\ 0.5812 & 0.7039 & -0.4082 \end{bmatrix}^T$$

The above method of representing the matrix A as the product of the Unitary matrix 'U_1', diagonal matrix 'D' and the Unitary matrix '$U_2{}'^T$ are called Singular Value Decomposition.

1.27 Applications of Singular Value Decomposition

1. Spectral factorization representation of the matrix is obtained using Singular Value Decomposition as given below.

   Example 1.27. From the Example 1.23, we get,

$$\begin{bmatrix} 1 & 2 & 3 \\ 4 & 5 & 6 \end{bmatrix}.$$

$$= \begin{bmatrix} 0.9224 & 0.3863 \\ -0.3863 & 0.9224 \end{bmatrix} \begin{bmatrix} \sqrt{0.5963} & 0 \\ 0 & \sqrt{90.4027} \end{bmatrix} \begin{bmatrix} -0.8060 & -0.4287 \\ -0.1124 & 0.5663 \\ 0.5812 & 0.7039 \end{bmatrix}^T$$

$$= \begin{bmatrix} 0.9224 & 0.3863 \\ -0.3863 & 0.9224 \end{bmatrix} \begin{bmatrix} \sqrt{0.5963} & 0 \\ 0 & \sqrt{90.4027} \end{bmatrix} \begin{bmatrix} -0.8060 & -0.1124 & 0.5812 \\ -0.4287 & 0.5663 & 0.7039 \end{bmatrix}$$

$$= \sqrt{0.5963}\begin{bmatrix} 0.9224 \\ -0.3863 \end{bmatrix}\begin{bmatrix} -0.8060 & -0.1124 & 0.5812 \end{bmatrix}$$

$$+ \sqrt{90.4027}\begin{bmatrix} 0.3863 \\ 0.9224 \end{bmatrix}\begin{bmatrix} -0.4287 & -0.5663 & 0.7039 \end{bmatrix}$$

The above mentioned way of representing the matrix is called spectral factorization. If the Eigen values are so small, we can neglect them in the above representation and hence data compression is achieved.

2. Computation of Pseudo inverse of the non-invertible matrix
Case 1: When the matrix A^TA is invertible
Consider the case of solving the equation of the form $Ax = b$ when A^TA is invertible.

Let the equation be $\begin{bmatrix} 1 & 0 & 0 \\ 0 & 2 & 0 \\ 0 & 0 & 3 \\ 0 & 0 & 0 \end{bmatrix}\begin{bmatrix} x1 \\ x2 \\ x3 \end{bmatrix} = \begin{bmatrix} 5 \\ 6 \\ 4 \\ 4 \end{bmatrix}$.

Multiplying A^T on both sides, we get

$$\begin{bmatrix} 1 & 0 & 0 & 0 \\ 0 & 2 & 0 & 0 \\ 0 & 0 & 3 & 0 \\ 0 & 0 & 0 & 0 \end{bmatrix}\begin{bmatrix} 1 & 0 & 0 \\ 0 & 2 & 0 \\ 0 & 0 & 3 \\ 0 & 0 & 0 \end{bmatrix}\begin{bmatrix} x1 \\ x2 \\ x3 \end{bmatrix} = \begin{bmatrix} 1 & 0 & 0 & 0 \\ 0 & 2 & 0 & 0 \\ 0 & 0 & 3 & 0 \\ 0 & 0 & 0 & 0 \end{bmatrix}\begin{bmatrix} 5 \\ 6 \\ 4 \\ 4 \end{bmatrix}$$

$$\Rightarrow \begin{bmatrix} 1 & 0 & 0 \\ 0 & 4 & 0 \\ 0 & 0 & 9 \end{bmatrix}\begin{bmatrix} x1 \\ x2 \\ x3 \end{bmatrix} = \begin{bmatrix} 5 \\ 12 \\ 12 \end{bmatrix}$$

In this case $x1 = 5, x2 = 3, x3 = \frac{4}{3}$

Note that the solution corresponds to the projected column vector $\begin{bmatrix} 1 & 0 & 0 \\ 0 & 2 & 0 \\ 0 & 0 & 3 \\ 0 & 0 & 0 \end{bmatrix}$

$$\begin{bmatrix} x1 \\ x2 \\ x3 \end{bmatrix} = \begin{bmatrix} 5 \\ 6 \\ 4 \\ 0 \end{bmatrix}$$

Also note that the matrix $A^T A$ is invertible and we have already shown that the solution vector thus obtained is such that magnitude of the error vector $\begin{bmatrix} 5 \\ 6 \\ 4 \\ 4 \end{bmatrix} - \begin{bmatrix} 5 \\ 6 \\ 4 \\ 0 \end{bmatrix} = \begin{bmatrix} 0 \\ 0 \\ 0 \\ 4 \end{bmatrix}$ is minimized.

Case 2: When the matrix A^TA is non-invertible
Consider the case of solving the equation of the form $Ax = b$, where $A^T A$ is non-invertible.

$$\text{Let the equation be} \begin{bmatrix} 1 & 0 & 0 & 0 \\ 0 & 2 & 0 & 0 \\ 0 & 0 & 3 & 0 \end{bmatrix} \begin{bmatrix} x1 \\ x2 \\ x3 \\ x4 \end{bmatrix} = \begin{bmatrix} 5 \\ 6 \\ 4 \end{bmatrix}.$$

In this case $A' * A = \begin{bmatrix} 1 & 0 & 0 & 0 \\ 0 & 4 & 0 & 0 \\ 0 & 0 & 9 & 0 \\ 0 & 0 & 0 & 0 \end{bmatrix}$ is not invertible and hence projection method as used above cannot be used in this case.
By direct observation of the set of equations, the solution is obtained as follows

$$x1 = 5$$
$$x2 = 3$$
$$x2 = 4/3$$

The variable $x4$ is arbitrarily chosen as 0 to reduce the length of the vector $\begin{bmatrix} x1 \\ x2 \\ x3 \\ x4 \end{bmatrix}$.
Thus the solution obtained above is of shortest length.

In both the cases the pseudo inverse matrix of the diagonal matrix A which is represented as A^+ is obtained as follows.

$$\text{For the Case 1 } A^+ = \begin{matrix} 1 & 0 & 0 & 0 \\ 0 & \frac{1}{2} & 0 & 0 \\ 0 & 0 & \frac{1}{3} & 0 \end{matrix}$$

The solution for the equation $Ax = b$, is obtained as $x = A^{+}b$

$$\Rightarrow x = \begin{bmatrix} 1 & 0 & 0 & 0 \\ 0 & \frac{1}{2} & 0 & 0 \\ 0 & 0 & \frac{1}{3} & 0 \end{bmatrix} \begin{bmatrix} 5 \\ 6 \\ 4 \end{bmatrix} = \begin{bmatrix} 5 \\ 3 \\ 4/3 \end{bmatrix}$$

Similarly for the case 2,

$$A^{+} = \begin{matrix} 1 & 0 & 0 \\ 0 & \frac{1}{2} & 0 \\ 0 & 0 & \frac{1}{3} \\ 0 & 0 & 0 \end{matrix}$$

The solution for the equation $Ax = b$, is obtained as $x = A^{+}b$

$$\Rightarrow x = \begin{bmatrix} 1 & 0 & 0 \\ 0 & \frac{1}{2} & 0 \\ 0 & 0 & \frac{1}{3} \\ 0 & 0 & 0 \end{bmatrix} \begin{bmatrix} 5 \\ 6 \\ 4 \end{bmatrix} = \begin{bmatrix} 5 \\ 3 \\ 4/3 \\ 0 \end{bmatrix}$$

Thus if the matrix A is the diagonal matrix the inverse of the matrix can be obtained by inverting the non-zero diagonal elements of the matrix A with zero elements unchanged.

If the matrix A is not the diagonal matrix, then SVD is used to represent the matrix as $A = U_1 D {U_2}^H$ and obtain the vector x such that $\|Ax - b\|$ is minimized.

$\Rightarrow$ we have to obtain the value for the vector x such that $\|U_1 D {U_2}^H x - b\|$ is minimized.
$\Rightarrow$ $\|D {U_2}^H x - {U_1}^H b\|$ is minimized. [Multiplying Unitary matrix on both sides of the linear equation]

Note:
Multiplying with the unitary matrix on both sides of the linear equation will not affect the distance as described below
In general $\|Ax - b\|^2 = (Ax - b)^H (Ax - b)$
Multiplying the Unitary matrix U on both sides we get,

$$\begin{aligned} \|UAx - Ub\|^2 &= (UAx - Ub)^H (UAx - Ub) = U \\ &= x^H A^H U^H UAx - x^H A^H U^H Ub - b^H U^H Ub \\ &\quad - b^H U^H Ub \\ &= x^H A^H Ax - x^H A^H b - b^H b - b^H b \\ &= (Ax - b)^H (Ax - b) = \|Ax - b\|^2 \end{aligned}$$

Let $y = U_2{}^H x$ and the modified task is to minimize the norm $\|Dy - U_1{}^H b\|$, where 'D' is the diagonal matrix and hence solution for y is obtained as

$$\begin{aligned} y &= D^+ U_1{}^H b \\ &\Rightarrow U_2{}^H x = D^+ U_1{}^H b \\ &\Rightarrow x = U_2 D^+ U_1{}^H b \end{aligned}$$

Thus for any arbitrary matrix A, we can obtain inverse (If the matrix is invertible) or pseudo inverse (If the matrix is not invertible) using the technique as mentioned above.

Example 1.28.

$$A = \begin{bmatrix} 1 & 2 & 3 \\ 4 & 5 & 6 \end{bmatrix}$$

The inverse of the matrix A is obtained as follows.
Using SVD we get,

$$\begin{bmatrix} 1 & 2 & 3 \\ 4 & 5 & 6 \end{bmatrix}$$

$$= \begin{bmatrix} 0.9224 & 0.3863 \\ -0.3863 & 0.9224 \end{bmatrix} \begin{bmatrix} \sqrt{0.5963} & 0 & 0 \\ 0 & \sqrt{90.4027} & 0 \\ 0 & 0 & 0 \end{bmatrix} \begin{bmatrix} -0.8060 & -0.4287 & 0.4082 \\ -0.1124 & 0.5663 & 0.8165 \\ 0.5812 & 0.7039 & -0.4082 \end{bmatrix}^T$$

A^+

$$= \begin{bmatrix} 0.9224 & 0.3863 \\ -0.3863 & 0.9224 \end{bmatrix} \begin{bmatrix} \frac{1}{\sqrt{0.5963}} & 0 & 0 \\ 0 & \frac{1}{\sqrt{90.4027}} & 0 \\ 0 & 0 & 0 \end{bmatrix} \begin{bmatrix} -0.8060 & -0.4287 & 0.4082 \\ -0.1124 & 0.5663 & 0.8165 \\ 0.5812 & 0.7039 & -0.4082 \end{bmatrix}^T$$

A^+

$$= \begin{bmatrix} 0.9224 & 0.3863 \\ -0.3863 & 0.9224 \end{bmatrix} \begin{bmatrix} \frac{1}{\sqrt{0.5963}} & 0 & 0 \\ 0 & \frac{1}{\sqrt{90.4027}} & 0 \\ 0 & 0 & 0 \end{bmatrix} \begin{bmatrix} -0.8060 & -0.4287 & 0.4082 \\ -0.1124 & 0.5663 & 0.8165 \\ 0.5812 & 0.7039 & -0.4082 \end{bmatrix}^T$$

3. Representing the matrix A as the product of Unitary matrix and the symmetric matrix as shown below.
Using SVD, $A = U_1 D U_2{}^H$
Inserting $U_2{}^H U_2$ in the middle we get

$$A = U_1 U_2{}^H U_2 D U_2{}^H$$

Note that the matrix $U_1 U_2{}^H$ is the unitary matrix and the matrix $U_2 D U_2{}^H$ is the symmetric matrix.

Chapter 2
Probability

2.1 Introduction

1. Set: It is collection of well defined objects. Each object is referred as an element
2. The set B which is the subset of A (Represented as $B \subset A$), is a set whose element are also the elements of A
3. Set of no elements are called Empty set
4. Set operations:
   - A union B (Represented as $A \cup B$ or $A + B$) is the set that consists of the elements which are either in A or B or in both.
   - A intersection B (Represented as $A \cap B$ or AB) is the set of elements which are in both A and B.
   - A complement (Represented as $\bar{A}$ or A^c) is the set that consists of the elements that are not present in the set A.
5. Mutually exclusive sets (Disjoint sets): Two sets A and B are disjoint, if they have no elements in common i.e. $AB = \phi$. In general The sets A1,A2,A3,... An are disjoint if Ai Aj $= \phi$ for $i \neq j$
6. Sample space: Set of all experimental outcomes
7. Partition: A partition of a set is the collection of mutually exclusive sets A1, A2...whose union is the sample space 'S'
   (i.e.) If $A1 + A2 + A3 + \ldots An = S$ and Ai Aj $= \phi$ for $i \neq j$, then Ai's form a partition of the set
8. Event: Subset of the sample space
   Certain event: S (Sample space)
   Impossible event: ϕ (Null space)
   Elementary event: One outcome of the experiment.
9. Countable infinite set: The set C is countably infinite when there is one-to-one correspondence between the elements of the set C and a set of all non-negative numbers

E.S. Gopi, *Mathematical Summary for Digital Signal Processing Applications with Matlab*, DOI 10.1007/978-90-481-3747-3_2,

2.2 Axioms of Probability

1. Probability of the event ξ represented as P(ξ) ≥ 0.
2. P(S) $= 1$, where 'S' is the sample space.
3. If A1,A2,... An are such that Ai Aj $= \phi$ for i $\neq$ j then

$$\mathrm{P}\left(\sum_{i=1}^{\infty} Ai\right) = \sum_{i=1}^{\infty} P(Ai).$$

Note: If sample space is having finite number or infinitely countable number of subsets, probabilities can be assigned to all the subsets of the sample space that satisfies all the axioms mentioned in 2.2.

If sample space is having uncountable infinite number of subsets, it is **not possible** to assign the probability to all the subsets of the sample space that satisfies all the axioms mentioned in 2.2.

2.3 Class of Events or Field (F)

Class of Events is the subset of the sample space satisfying the following properties.

1. If A $\subset$ F, then $\bar{A} \subset$ F.
2. If A, B $\in$ F, then A $+$ B $\in$ F.
3. Sigma Field: If A1, A2, ... $\in$ F, $\sum_{i=1}^{\infty} Ai \in$ F.

2.4 Probability Space (S, F, P)

The probability space (S, F, P) consists of Sample space S, Field F and the probability measure P. The probability measure maps every element of F to a number less than 1 and greater than 0. The measured value is the probability.

2.5 Probability Measure

The probability of the event A can be measured as follows.
Method 1:

Probabilities of elementary events are assumed as equal. Let the number of outcomes belonging to the event A is N_A and the total number of outcomes is N, then P(A) $= N_A/N$. The probability measured in this technique depends upon how the set of possible outcomes are defined.

Method 2:

Repeat the experiment N times. Let n_A is the number of times event A occurs, then P(A) = $\lim_{N\to\infty} \frac{n_A}{N}$

2.6 Conditional Probability

Case 1 (Fig. 2.1):

Let 'nA' and 'nB' be the number of outcomes belonging to the event A and B respectively. Also let 'n' be the total number of outcomes.

Probability of the event A is nA/n

Probability of the event B is nB/n

Probability of the event A given B has occurred is given as follows

$$P(A/B) = nA/nB = \frac{nA}{n} \Big/ \frac{nB}{n} = P(A)/P(B) = P(AB)/P(B)$$

Case 2 (Fig. 2.2):

In this case P(A/B) = Probability of A given B has occurred is 1. This can also be obtained using the formula obtained in the case 1. P (A/B) = P (AB)/P(B) = P(B)/P(B) = 1

Therefore P(A/B) = P(AB)/P(B) is considered as the common formula that can be used in both the cases.

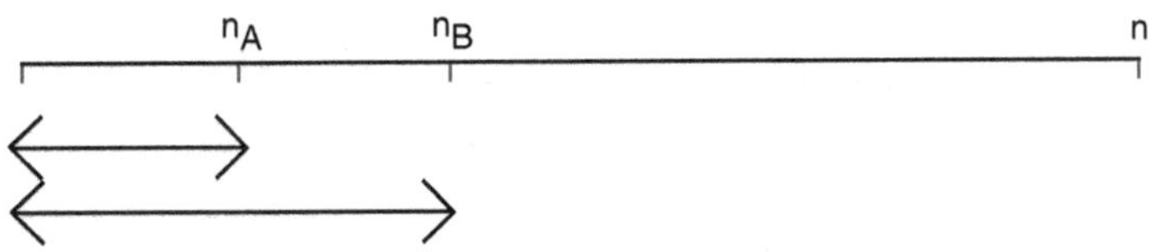

Fig. 2.1 Conditional probability – case 1

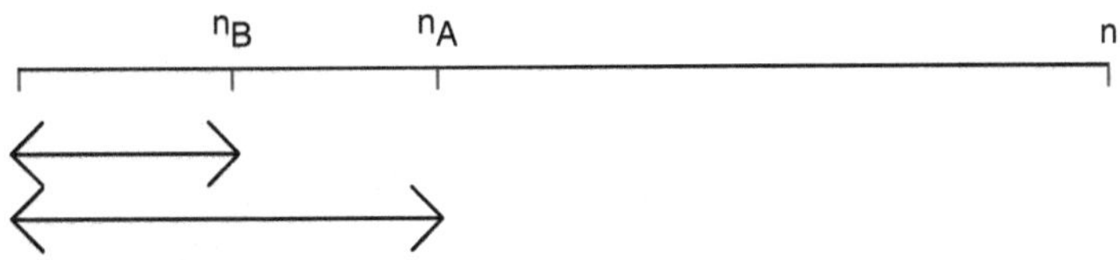

Fig. 2.2 Conditional Probability – case 2

2.7 Total Probability Theorem

Let A1, A2 ... An be the partition of the sample space S.

$$B = BS = B(A1 + A2 + A3 + \ldots An) = A1B + A2B + A3B + \ldots AnB$$

$$P(B) = \sum\nolimits_{i=1}^{n} P(AiB)$$

$$= \sum_{i=1}^{n} P\left(\frac{B}{Ai}\right) P(Ai)$$

$$\textit{Total Probability Theorm}: P(B) = \sum_{i=1}^{n} P\left(\frac{B}{Ai}\right) P(Ai)$$

Note that Ai Aj = ϕ for i $\neq$ j

2.8 Bayes Theorem

$$P(A/B) = P(AB)/P(B)$$

$$P(B/A) = P(AB)/P(A)$$

$$P(B)P(A/B) = P(B/A)P(A)$$

$$\Rightarrow P(B) = P(A)P(B/A)/P(A/B)$$

$$\textit{Bayes Theorem}: \; P(B) = P(A)\, P\left(\frac{B}{A}\right) \Big/ P\left(\frac{A}{B}\right)$$

Let A1, A2, ... An forms the partition of the sample space S.

In general $P(Ai/B) = \frac{P(Ai)P(\frac{B}{Ai})}{\sum_{i=1}^{n} P(\frac{B}{Ai})P(Ai)}$

Note that Ai Aj = ϕ for i $\neq$ j

2.9 Independence

1. A and B are independent if P(AB) = P(A)P(B). Also P(A/B) = P(A)
2. Three events A, B and C are independent if it satisfies the following conditions

$$P(AB) = P(A)P(B)$$
$$P(AC) = P(A)P(C)$$
$$P(BC) = P(B)P(C)$$
$$P(ABC) = P(A)P(B)P(C)$$

$$P(AB/C) = P(ABC)/P(C)$$
$$P(A/BC) = \frac{P(ABC)}{P(BC)} = P(A)$$

2.10 Multiple Experiments (Combined Experiments)

1. Consider the sample spaces S1 and S2 corresponding to the two independent experiments.

$$\{\mathrm{S1}\} = \{1, 2, 3, 4, 5, 6\}$$
$$\{\mathrm{S2}\} = \{\mathrm{h}, \mathrm{t}\}$$

   Sample space of the combined experiments is represented as S = S1XS2 = {(a, b), a ∈ S1, b ∈ S2}
2. Consider the event A = {2, 3} ∈ S1 and the event B = {h} ∈ S2.

$$\text{The event}\{\mathrm{AXS2}\} = \{(2, h)(3, h)(2, t)(3, t)\}$$
$$\text{The event}\{\mathrm{S1XB}\} = \{(1, h)(2, h)(3, h)(4, h)(5, h)(6, h)\}$$
$$\{\mathrm{AXB}\} = \{(2, h)(3, h)\}\text{which can be obtained using the following.}$$
$$\{\mathrm{AXB}\} = \{\mathrm{AXS2}\} \cap \{\mathrm{S1XB}\}$$
$$\mathrm{P(AXB)} = \mathrm{P(AXS2)P(S1XB)}$$

Some properties of the probability derived from the Axioms of the probability

1. $P(\bar{A}) = 1 - P(A)$

   Proof. $A + \bar{A} = S$ *(A and $\bar{A}$ are Mutually Exclusive Events)*

$$P(S) = P(A + \bar{A}) = 1 \text{ [\textit{Second Axiom}]}$$
$$= P(A) + P(\bar{A}) \text{ [\textit{Third Axiom}]}$$
$$\Rightarrow P(A) = 1 - P(\bar{A})$$

2. $P(A) \leq 1$

   Proof. $P(A) = 1 - P(\bar{A})$ *[From proof 1]*

$$P(\bar{A}) \geq 0 \textit{ and } P(A) \geq 0 \text{ [\textit{First Axiom}]}$$
$$\Rightarrow P(A) \leq 1$$

3. $P(\Phi) = 0$

   Proof. *The Event* $A = A + \Phi$ *[A and Φ are Mutually Exclusive Events]*

$$P(A) = P(A) + P(\Phi) \text{ [\textit{Third Axiom}]}$$
$$\Rightarrow P(\Phi) = 0$$

4. $\{A_i\}_{i=1}^{n}$ *is a partition of the sample space* S (Fig. 2.3). *For any event* $B, P(B) = \sum_{i=1}^{n} P(B\ A_i)$

Proof. From the diagram, it can be shown that any event 'B' can be represented as the union of disjoint events $\bigcup_{i=1}^{n} B\ A_i$

$$
\begin{aligned}
(i.e.)\ B &= \cup_{i=1}^{n} BA_i \\
P(B) &= P(\cup_{i=1}^{n} B\ A_i) = P(BA_1 + BA_2 + BA_3 + BA_4 + \cdots BA_n) \\
&= P(BA_1) + P(BA_2) + \ldots P(BA_n) = \sum_{i=1}^{n} P(BA_i)
\end{aligned}
$$

5. *If* $A \subset B,\ P(A) \leq P(B)$ (Fig. 2.4)

Proof. $A = B + \bar{B}A$ *(Note that B and* $\bar{B}A$ *are disjoint events)*

$$\Rightarrow P(A) = P(B) + P(\bar{B}A)$$

$$Also\ 1 \geq P(\bar{B}A) \geq 0\ and\ hence\ P(A) \geq P(\bar{B}A)$$

6. $P(A+B+C) = P(A)+P(B)+P(C)-P(AB)-P(AC) = P(BC)+P(ABC)$ (Fig. 2.5)

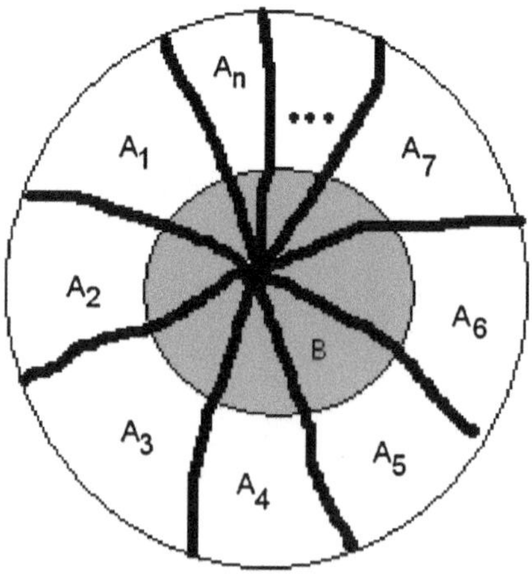

Fig. 2.3 Partition of the sample space S

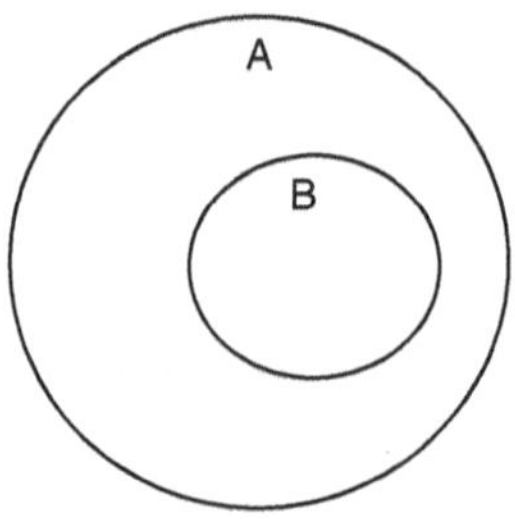

Fig. 2.4 Venn diagram illustrating $A \subset B$

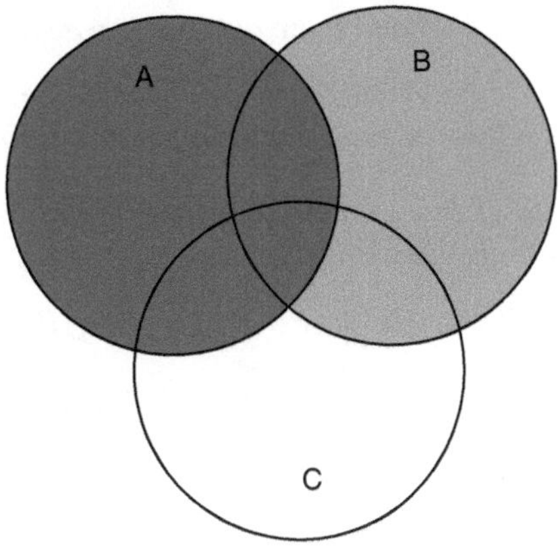

Fig. 2.5 Venn diagram illustrating $A + B + C$

Proof. $A + B + C$ *can be represented as the summation of three disjoint events as given below.*

$$A + B + C = A + \bar{A}B + \overline{(A + B)}C \text{ (Disjoint Events)}$$
$$\Rightarrow A + B + C = A + (1 - A)B + \bar{A}\bar{B}C$$
$$\Rightarrow P(A + B + C) = P(A) + P((1 - A)B) + P(\bar{A}\bar{B}C)$$

On simplification

$$A + B + C = A + (1 - A)B + (1 - A)(1 - B)C \text{ (Disjoint Events)}$$
$$\Rightarrow A + B + C = A + B - AB + (1 - B - A + AB)C$$
$$\Rightarrow A + B + C = A + B - AB + C - BC - AC + ABC$$
$$\Rightarrow P(A + B + C) = P(A) + P(B) - P(AB) + P(C) - P(BC)$$
$$-P(AC) + P(ABC)$$
$$\Rightarrow P(A + B + C) \leq P(A) + P(B) + P(C)$$

7. *In general* $P\left(\{A_i\}_{i=1}^n\right) \leq \sum_{i=1}^n P(A_i)$
   (It can be proved using Mathematical Induction)
8. *If two events A and B are independent,* $\bar{A}$ *and B are independent*

Proof. *Given*: $P(AB) = P(A)P(B)$

$$\bar{A}B = B - AB$$
$$P(\bar{A}B) = P(B) - P(AB)$$
$$\Rightarrow P(B) - P(A)P(B)$$
$$\Rightarrow P(B)(1 - P(A))$$
$$\Rightarrow P(B)P(\bar{A})$$

Thus $\bar{A}$ and B are also independent.

$\Rightarrow$ *$\bar{A}$ and $\bar{B}$ are independent* ***(property 8)***
$\Rightarrow$ *A and $\bar{B}$ are independent* ***(property 8)***

9. *If three events A, B and C are independent, the events A and B + C are independent.*

Proof. *Given*: $P(ABC) = P(A)P(B)P(C)$

$$\begin{aligned} P(A(B+C)) = P(AB+AC) &= P(AB) + P(AC) - P(ABC) \\ &= P(A)P(B) + P(A)P(C) - P(A)P(B)P(C) \\ &= P(A)(P(B) + P(C) - P(BC)) \\ &= P(A)P(B+C) \\ &\Rightarrow A \text{ and } B + C \text{ are independent} \end{aligned}$$

10. *If A, B and C are three events,* $P(AB/C) = P(A/BC)P(B/C)$.

Proof.

$$\begin{aligned} P(AB/C) &= P(ABC)/P(C) \\ &= P(A/BC)P(BC)/P(C) \\ &= P(A/BC)P(B/C) \end{aligned}$$

11. If A, B and C are three events, P(ABC) = P(A/BC)P(B/C)P(C).

Proof.

$$\begin{aligned} \text{P(ABC)} &= \text{P(A/BC)P(BC)} \\ &= \text{P(A/BC)P(B/C)P(C)} \end{aligned}$$

2.11 Random Variable

Consider the probability space (S,F,P), then the mapping of the outcomes s $\in$ F to the real line is called random variable. (i.e.) X: F$-$ $>$ R (Fig. 2.6).

Mapping must be chosen such that every subset of the real line of the form $(-\infty$, a] should be an event in F. The subset of the real line of the form (a, b] is called Borel set represented as B.

Let A $\in$ B. P_X(A) = P({s $\in$ S : X(s) $\in$ A}), where {s $\in$ S : X(s) $\in$ A} is called inverse image of A. Thus the random variable maps the probability space (S,F,P) to the Borel space (R, B, Px)

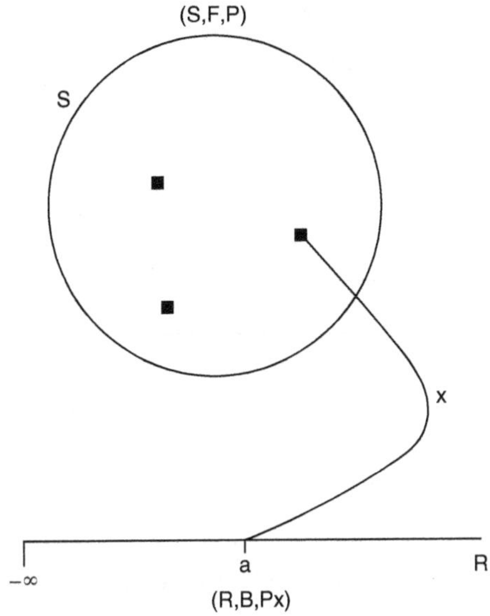

Fig. 2.6 Illustration of the random variable

2.12 Cumulative Distribution Function (cdf) of the Random Variable 'x'

$$F_X(\alpha) = P_X(X \leq \alpha)$$

1. $0 \leq F_X(\alpha) \leq 1$ *for all* α
2. $\lim_{\alpha \to \infty} F_X(\alpha) = 1$
3. $\lim_{\alpha \to -\infty} F_X(\alpha) = 0$
4. $F_X(\alpha)$ is the non decreasing function. (i.e.) If $\alpha 1 < \alpha 2$, then $F_X(\alpha 1) \leq F_X(\alpha 2)$
5. If $F_X(\alpha 0) = 0$, *then* $F_X(\alpha) = 0$ for all $\alpha < \alpha 0$
6. $P_X(a \leq X \leq b) = F_X(b) - F_X(a)$
7. If X and Y are two random variable such that $X(s) \leq Y(s)$ for all s $\in$ S, then $F_X(a) \geq F_Y(a)$ for all α (Fig. 2.7)

Proof.

$$F_X(a) = P(X \leq a)$$
$$F_Y(b) = P(Y \leq b)$$

By definition

$$\begin{aligned} F_X(a) &= F_Y(b) \\ &= P(Y \leq a) + P(a \leq Y \leq b) \\ &= F_Y(a) + \mathrm{K} \end{aligned}$$

The constant K ranges from 0 to 1 and it is the positive quantity and hence $F_X(a) \geq F_Y(a)$ for all values of 'a'

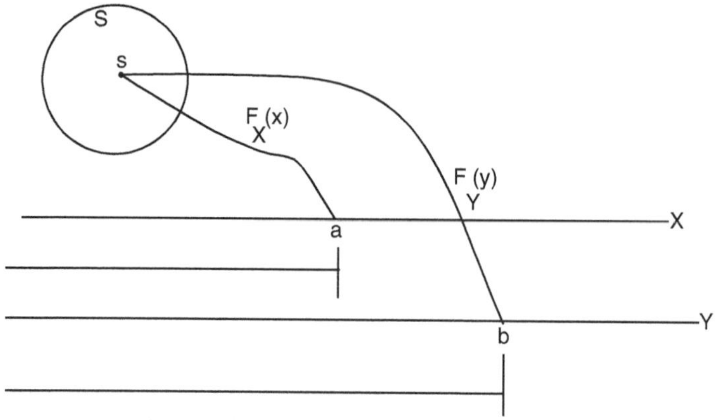

Fig. 2.7 Illustration of the property 7 of the CDF

2.13 Continuous Random Variable

A random variable X is continuous if $F_X(\alpha)$ is the continuous function of α.

2.14 Discrete Random Variable

A random variable X is discrete if and only if it maps S to a countable subset of R (Real numbers).

2.15 Probability Mass Function

If the random variable is discrete, probability of the particular value of the random variable can be computed using the function known as probability mass function. Probability of the random variable X = α is represented as P(X = α).

2.16 Probability Density Function

If the random variable is continuous, probability of the random variable X over the smallest range α *to* $\alpha + \Delta\alpha$ is computed as follows.

$$\lim_{\Delta\alpha \to 0} [F_X(\alpha + \Delta\alpha) - F_X(\alpha)]$$

Define $f_x(\alpha)\Delta\alpha$ be the probability of the random variable x over the smallest range α *to* $\alpha + \Delta\alpha$

$$\Rightarrow f_x(\alpha)\Delta\alpha = \lim_{\Delta\alpha\to 0}[F_X(\alpha+\Delta\alpha) - F_X(\alpha)]$$

$$\Rightarrow f_x(\alpha) = \lim_{\Delta\alpha\to 0}\,[F_X(\alpha+\Delta\alpha) - F_X(\alpha)]\,/\Delta\alpha$$

The function $f_x(\alpha)$ is called probability density function.

$$f_x(\alpha)\frac{\lim\limits_{\Delta\alpha\to 0}[F_X(\alpha+\Delta\alpha) - F_X(\alpha)]}{\Delta\alpha}$$

$$\Rightarrow f_x(\alpha) = \frac{dF_X(\alpha)}{dx}$$

Properties:

1. $f_x(\alpha) \geq 0$ for all α. F [Because F is the non-decreasing function]
2. $F_X(\alpha) = \int_{-\infty}^{\alpha} f_x(\alpha)\,d\alpha$
3. $\int_{-\infty}^{\infty} f_x(x)dx = 1$
4. $\int_{x1}^{x2} f_x(x)dx = F_X(x2) - F_X(x1)$

2.17 Two Random Variables

Consider the probability space (S, F, P), then the mapping of the outcome s $\in$ F to the real line is called random variable. This mapping can be done in multiple forms and are called as Multiple random variables. In particular if the mapping is done in two different forms, then the corresponding mapping is called two random variables as shown in the Fig. 2.8. Two random variables are completely described by the Joint distribution function $P[X \leq x,\ Y \leq y]$ and it is usually represented as $F_{XY}(x,\ y)$.

Properties of joint distribution function and joint density function with two random variables.

1. $\lim_{x\to\infty} F_{XY}(x, y) = P(X \leq \infty,\ Y \leq y) = F_Y(y)$.
2. $\lim_{y\to\infty} F_{XY}(x, y) = P(X \leq x,\ Y \leq \infty) = F_X(x)$.
3. $\lim_{Y\to-\infty} F_{XY}(x, y) = 0$
4. $\lim_{X\to-\infty} F_{XY}(x, y) = 0$
5. $P(x1 \leq X \leq x2, Y \leq \infty) = F_{XY}(x2, y) - F_{XY}(x1, y)$.
6. $P(x1 \leq X \leq x2, y1 \leq Y \leq y2) = F_{XY}(x2, y2) - F_{XY}(x2, y1) - F_{XY}(x1, y2) + F_{XY}(x1, y1)$
7. $0 \leq F_{XY}(x, y) \leq 1$
8. It is the non-decreasing function with both 'x' and 'y'

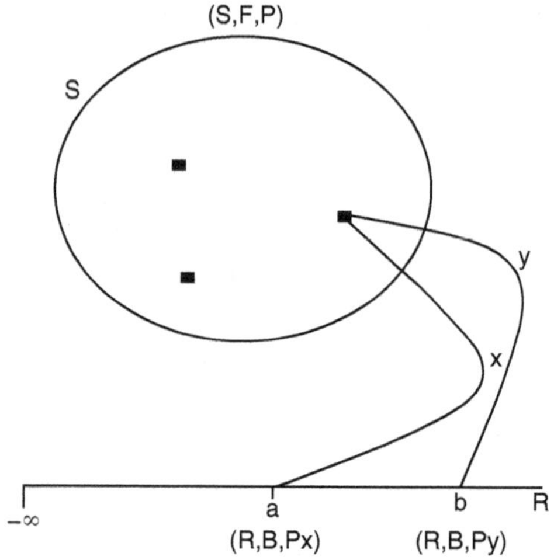

Fig. 2.8 Illustration of the two random variables

9. The relationship between the Joint probability density function of the two random variables 'X' and 'Y' and the corresponding joint distribution function is given below

$$f_{XY}(x,y) = \frac{\partial^2 \mathrm{F_{XY}(x,y)}}{\partial x \partial y}$$

$$\mathrm{F_{XY}(a,b)} = \int_{-\infty}^{b} \int_{-\infty}^{a} f_{XY}(x,y)dx\, dy$$

10. Probability of the event A corresponding to the random variable X which ranges from xmin to xmax and the random variable Y which ranges from ymin to ymax is computed using the joint density function as given below

$$P(A) \int_{xmin}^{xmax} \int_{ymin}^{ymax} f_{XY}(x,y)dx\, dy$$

11. The marginal probability density function of the random variable 'X' is given as (Fig. 2.9)

$$f_X(x) = \int_{-\infty}^{\infty} f_{XY}(x,y)\, dy$$

$$F_Y(y) = \int_{-\infty}^{\infty} f_{XY}(x,y)\, dx$$

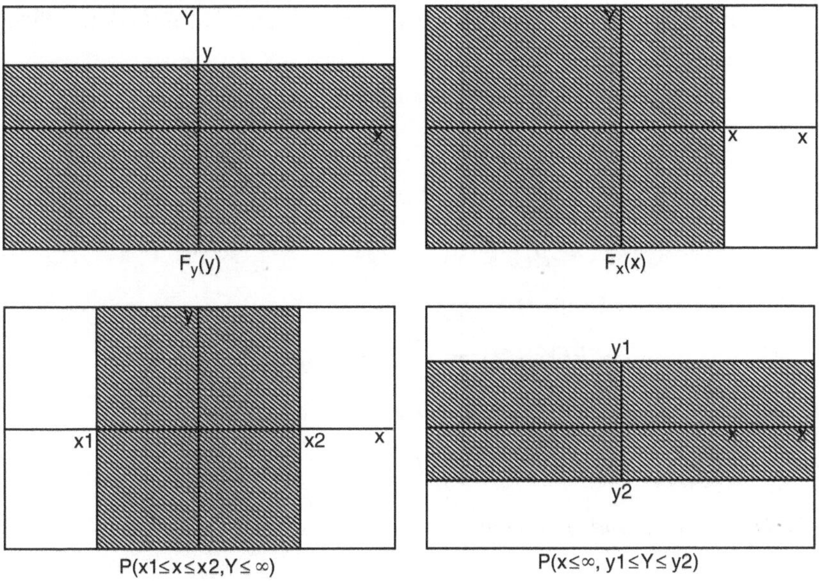

Fig. 2.9 Illustration of the Marginal probability density function

2.18 Conditional Distributions and Densities

The conditional distribution of 'x' over the event 'B' represented as

$$\mathrm{F}_{X/B}(\mathrm{x}) = \mathrm{P}(X \leq x/B) = \frac{\mathrm{P}(X \leq x, B)}{P(B)}$$

2.19 Independent Random Variables

Two events A and B are called independent if;

$$P(AB) = P(A)P(B)$$

Then we can write, $F_{XY}(x, y) = P(X \leq x, Y \leq y)$

$$= P(X \leq x)(Y \leq y)$$
$$= F_X(x)F_Y(y)$$

So, $f_{XY}(x, y) = f_X(x)f_Y(y)$

2.20 Some Important Results on Conditional Density Function

1. *'X' is continuous, 'Y' is discrete, then $F_{X/Y=y}(x)$ is computed as described below*

$$F_{X/Y=y}(x) = P(X \leq x/Y = y)$$
$$= \frac{P(X \leq x, Y = y)}{P(Y = y)}$$
$$P(X \leq x) = \sum_{Y} P(X \leq x, Y = y)$$
$$\Rightarrow P(X \leq x) = \sum_{Y} F_{X/Y=y}(x)P(Y = y)$$
$$\Rightarrow F_X(x) = \sum_{Y} F_{X/Y=y}(x)P(Y = y)$$
$$\Rightarrow f_X(x) = \sum_{Y} f_{X/Y=y}(x)P(Y = y)$$

2. *'X' is continuous and 'Y' is continuous, then $F_{X/Y=y}(x)$ is computed as follows*

$$F_{X/Y=y}(x) = P(X \leq x/Y = y)$$
$$= \lim_{\Delta y \to 0} P(X \leq x/y \leq Y \leq y + \Delta y)$$
$$= \lim_{\Delta y \to 0} \frac{P(X \leq x, y \leq Y \leq y + \Delta y)}{P(y \leq Y \leq y + \Delta y)}$$
$$= \lim_{\Delta y \to 0} \frac{P(y \leq Y \leq y + \Delta y/X \leq x)P(X \leq x)}{P(y \leq Y \leq y + \Delta y)}$$
$$= \lim_{\Delta y \to 0} \frac{P(y \leq Y \leq y + \Delta y/X \leq x)P(X \leq x)}{P(y \leq Y \leq y + \Delta y)}$$
$$= \lim_{\Delta y \to 0} \frac{(F_{Y/X \leq x}(y + \Delta y) - F_{Y/X \leq x}(y))P(X \leq x)}{F_Y(y + \Delta y) - F_Y(y + \Delta y)}$$
$$= \frac{(f_{Y/X \leq x}(y))P(X \leq x)}{f_Y(y)}$$
$$\Rightarrow P(X \leq x/Y = y) = \frac{(f_{Y/X \leq x}(y)P(X \leq x))}{f_Y(y)}$$

Also $P(X \leq x/Y = y)f_Y(y) = f_{Y/X \leq x}(y)P(X \leq x)$

$$\Rightarrow \int P(X \leq x/Y = y)f_Y(y)dy$$
$$= \int f_{Y/X \leq x}(y)P(X \leq x)dy$$
$$= P(X \leq x)$$

Also consider

$$P(X \le x/Y = y) = \lim_{\Delta y \to 0} \frac{P(X \le x, y \le Y \le y + \Delta y)}{P(y \le Y \le y + \Delta y)}$$

$$\Rightarrow F_{X/Y=y}(x) = \lim_{\Delta y \to 0} \frac{F_{XY}(x, y + \Delta y) - F_{XY}(x, y)}{F_Y(y) - F_Y(y)}$$

$$\Rightarrow F_{X/Y=y}(x) = \lim_{\Delta y \to 0} \frac{[F_{XY}(x, y + \Delta y) - F_{XY}(x, y)]/\Delta y}{[F_Y(y + \Delta y) - F_Y(y)]/\Delta y}$$

$$\Rightarrow F_{X/Y=y}(x) = \frac{\frac{dF_{XY}(x,y)}{dy}}{f_Y(y)}$$

$$\Rightarrow f_{X/Y=y}(x) = \frac{\frac{\partial^2 F_{XY}(x,y)}{\partial x \partial y}}{f_Y(y)}$$

$$\Rightarrow f_{X/Y=y}(x) = \frac{f_{XY}(x, y)}{f_Y(y)}$$

$$\Rightarrow \int f_{X/Y=y}(x) f_Y(y) dy = f_X(x)$$

$$\Rightarrow f_X(x) = \int f_{X/Y=y}(x) f_Y(y) dy$$

Example 2.1. Let $F_{XY}(x, y) =$

1	$x \in [2, \infty] \cap y \in [3, \infty]$
$^1/_2$	$x \in [0, 2] \cap y \in [3, \infty]$
$^1/_2$	$x \in [2, \infty] \cap y \in [0, 3]$
$^1/_4$	$x \in [0, 2] \cap y \in [0, 3]$
0	Else where

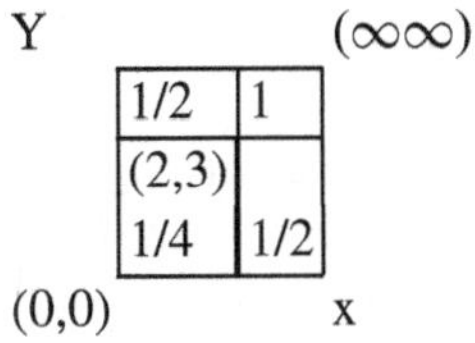

For the above specification $F_X(x)$ and $F_Y(y)$ are computed as follows (Fig. 2.10)

$$F_X(x) = P(X \le x, Y \le \infty) \; and \; F_Y(y) = P(X \le \infty, Y \le y)$$

Example 2.2. Let the event 'B' be $X \le b$

Then,

$$F_{X/X \le b}(x) = P(X \le x/X \le b) = \frac{P(X \le x, X \le b)}{P(X \le b)}$$

$$\Rightarrow F_{X/X \le b}(x) = 1 \qquad \text{for } X \ge b$$

$$= \frac{F_X(x)}{F_X(b)} \text{ for } X < b$$

Conditional Probability density function can be obtained by differentiating conditional distribution function.

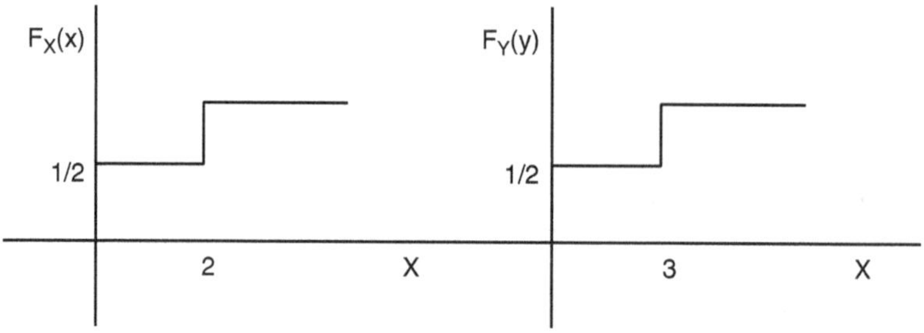

Fig. 2.10 $F_X(x)$ and $F_Y(y)$ of the Example 2.1

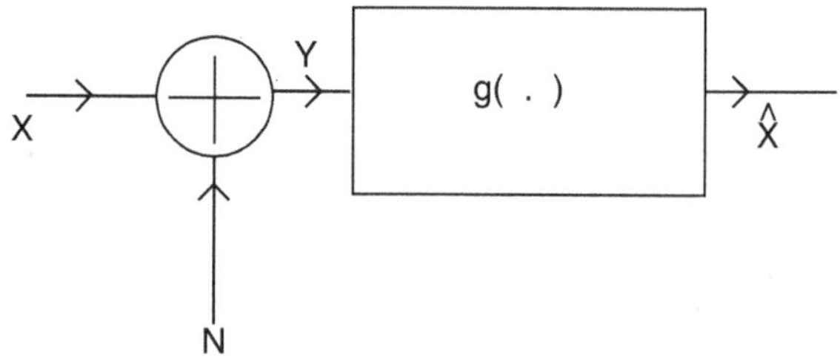

Fig. 2.11 Additive noise model of the channel transmission

$$\text{(i.e) } f_{X/B}(x) = \frac{dF_{X/B}(x)}{dx}$$

$$\Rightarrow f_{X/X\leq b}(x) = 0 \quad \text{for } X \geq b$$

$$= \frac{f_X(x)}{F_X(b)} \quad \text{for } X < b$$

Example 2.3. Consider the signal (random variable X) which holds the values 1 or −1 with known probability. Consider that signal that is corrupted by the additive noise signal (random variable N) to obtain the output signal (random variable Y). The probability density function of the random variable N is known to be Gaussian with mean = 0 (Fig. 2.11).

The corrupted signal Y is processed using the transformation function g (.) to estimate the value of the X. Consider the task of obtaining the transformation function g(.) so that the probability of correct decision is maximized.

$$\text{P(Correct decision)} = \int_Y P((correct\ decision)/Y = y) f_Y(y) dy$$

(See Section 2.20)

Let us redefine the problem as

$$\max .\ P((correct\ decision)/Y = y))\ \forall\ y$$

Suppose X = 1 is sent then,

$$P(X = 1/Y = y) > P(X = -1/Y = y)$$

$$\frac{f_{\frac{Y}{X}=1}(y)}{f_Y(y)}P(X=1) > \frac{f_{\frac{Y}{X}=-1}(y)}{f_Y(y)}P(X=-1)$$

$$\Rightarrow \frac{f_{Y/X=1}(y)}{f_{Y/X=-1}(y)} > \frac{P(X=-1)}{P(X=1)}$$

We have,

$$\begin{aligned} F_{Y/X=1}(y) &= P(Y \le y/X=1) \\ &= P(X+N \le y/X=1) \\ &= P(1+N \le y) \\ &= P(N \le y-1) \\ &= F_N(y-1) \end{aligned}$$

Differentiating both with respect to y,

$$\frac{\partial(F_{Y/X=1}(y)}{\partial y} = \frac{\partial F_N(y-1)}{\partial y}$$
$$f_{Y/X=1}(y) = f_N(y-1)$$

So,

$$f_{Y/X=1}(y) = f_N(y-1) \text{ and,}$$
$$f_{Y/X=-1}(y) = f_N(y+1)$$

Then,

$$\frac{f_N(y-1)}{f_N(y+1)} > \frac{P(X=-1)}{P(X=1)}$$

If $f_N(n)$ is a Gaussian density function with $\mu = 0$

$$f_N(n) = \frac{1}{\sqrt{2\pi\,\sigma^2}} e^{-(n)/2\sigma^2}$$

Then,

$$f_N(y-1) = \frac{1}{\sqrt{2\pi\,\sigma^2}} e^{-\frac{(y-1)^2}{2\sigma^2}}$$
$$f_N(y+1) = \frac{1}{\sqrt{2\pi\,\sigma^2}} e^{-\frac{(y+1)^2}{2\sigma^2}}.$$

Now,

$$\frac{F_N(y-1)}{f_N(y+1)} > \frac{P(X=-1)}{P(X=1)}$$

$$e^{-\frac{(y-1)^2}{2\sigma^2}} e^{\frac{(y+1)^2}{2\sigma^2}} > \frac{P(X=-1)}{P(X=1)}$$

$$e^{\frac{4y}{2\sigma^2}} > \frac{P(X=-1)}{P(X=1)}$$

$$\frac{4y}{2\sigma^2} > ln\frac{P(X=-1)}{P(X=1)}$$

$$y > \frac{\sigma^2}{2} ln\frac{P(X=-1)}{P(X=1)}$$

Thus the transformation function g(.) is obtained as follows.

$$If\ y > \frac{\sigma^2}{2} ln\frac{P(X=-1)}{P(X=1)} Decide\ \hat{X} = 1, otherside\ decide\ \hat{X} = -1$$

2.21 Transformation of Random Variables of the Type Y = g(X)

Consider the random variable 'X' which is transformed into another random variable 'Y' using the transformation function defined as Y = g(X). The graph relating the Y and X is as shown in the figure given below (Fig. 2.12).

Solving Y = g(X) for the general variable 'y1' gives x1, x4, x5. (i.e.) g(x1) = g(x4) = g(x5) = y1. Also note that they belongs to the region where there are piece-wise monotonically increasing or decreasing function.

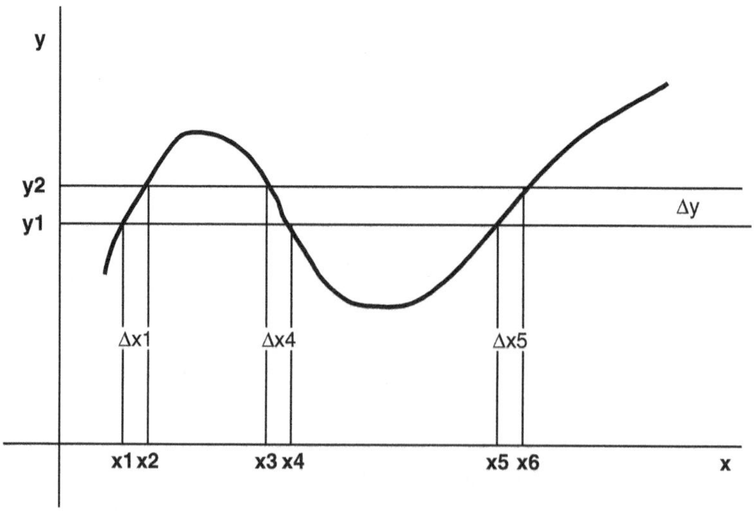

Fig. 2.12 Transformation of random variables-case 1

From the basics of probability theory

$$P(y1 \le y \le y2) = P(x1 \le x \le x2) + P(x3 \le x \le x4) + P(x5 \le x \le x6)$$
$$\Delta y \; f_Y(y) = \Delta x1 \; f_X(x) \; at \; x = x1 + \Delta x4 \; f_X(x)$$
$$= x4 + \Delta x5 \; f_X(x) \; at \; x = x5$$

Note that magnitude of $\Delta x1$, $\Delta x4$ and $\Delta x5$ are considered to compute $f_Y(y)$

$$\text{Therefore, } f_Y(y) = f_X(x1)\left|\frac{\Delta x1}{\Delta y}\right| + f_X(x4)\left|\frac{\Delta x4}{\Delta y}\right| + f_X(x5)\left|\frac{\Delta x5}{\Delta y}\right|$$

$$\Rightarrow f_Y(y) = \frac{f_X(x1)}{\left|\frac{dy}{dx}\right| at \; x = x1} + \frac{f_X(x4)}{\left|\frac{dy}{dx}\right| at \; x = x4} + \frac{f_X(x5)}{\left|\frac{dy}{dx}\right| at \; x = x5}$$

In general, probability density function of y (i.e.) $f_Y(y)$ is obtained as follows.
Given $f_X(x)$ and the transformation Y = g(X).

1. Obtain the solutions for the equation Y = g(x) for the general variable 'y', so that x1, x2, ... xn are obtained. Note that x1, x2, ... xn are represented in terms of 'y'.
2. $f_Y(y) = \sum_{i=1}^{n} \frac{f_X(xi)}{\left|\frac{dy}{dx}\right| at \; x=xi}$
3. The range of 'Y' can be obtained from the range of 'X' and the transformation equation Y = g(X).

2.22 Transformation of Random Variables of the Type Y1 = g1(X1, X2), Y2 = g2(X1, X2)

Given $f_{X1X2}(x1, x2), Y1 = g1(X1, X2), Y2 = g2(X1, X2), f_{Y1Y2}(y1, y2)$ is obtained as follows

Also note that the functions g1(.) and g2(.) are invertible. (i.e.) There exists the function h1(.) and h2(.) such that X1 = h1(Y1,Y2) and X2 = h2(Y1,Y2).

Consider the rectangle obtained from the intersection of lines Y1 = y1, Y1 = y1 + Δy1 and Y2 = y2, Y2 = y2 + Δy2 in the Y1–Y2 plane. The obtained rectangular region is represented as black shade in the Y1–Y2 plane.

Also, consider the curve 'a' in the X1–X2 plane which is obtained by joining the set of points (X1 = x1, X2 = x2) which satisfies the equation g1(x1, x2) = y1. Similarly curve 'b', curve 'c' and curve 'd' are obtained by joining the set of points (X1 = x1, Y1 = y1) which satisfies the equation g1(x1, x2) = y1 + Δy1, g2(x1, x2) = y2, g2(x1, x2) = y2 + Δy2 respectively. The shaded region represented in the X1–X2 plane (see Fig. 2.13)) is obtained as the intersection of the curves 'a', 'b', 'c' and 'd'. This shaded region can be approximated as the parallelogram.

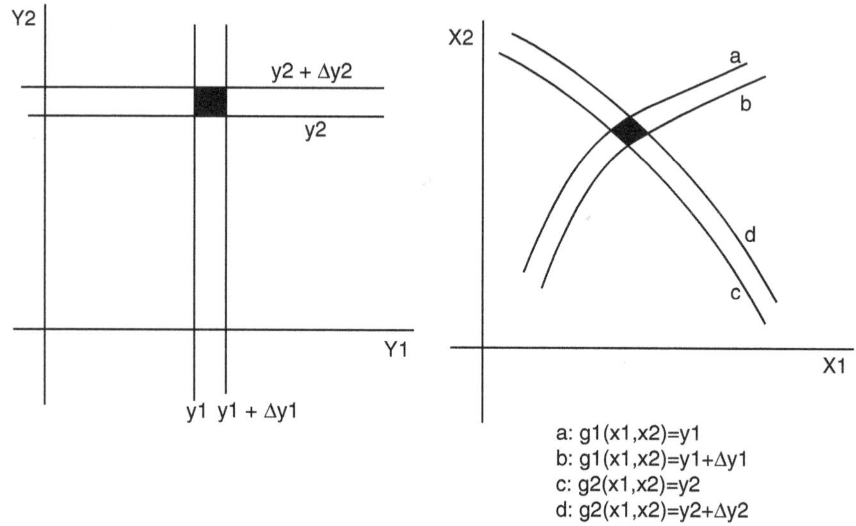

Fig. 2.13 Transformation of random variable-Case 2

From the basics of probability theory, the probability belonging to the shaded region in the Y1–Y2 plane is equal to the probability belonging to the shaded region in the X1–X2 plane. Therefore, $f_{Y1Y2}(y1, y2)$ [Area of the rectangle] $= f_{X1X2}(x1, x2)$ [Area of the area of the parallelogram]

Consider the curve a: $\mathrm{g1(x1,\ x2)} = \mathbf{y1}$. The set of points on this curve satisfies the equation $\mathrm{x1} = \mathrm{h1}(\mathbf{y1}, \mathrm{y2})$ and $\mathrm{x2} = \mathrm{h2}(\mathbf{y1}, \mathrm{y2})$. Note that '**y1**' is constant. Hence the points on this curve 'a' can be rewritten as $\mathrm{x1} = \mathrm{h1(y2)}$ and $\mathrm{x2} = \mathrm{h2(y2)}$ for the fixed $\mathrm{Y1} = \mathbf{y1}$.

Similarly for the curve b: The invertible equations for the equation $\mathrm{g1(x1,\ x2)} = \mathbf{y1} + \mathbf{\Delta y1}$ can be written as $\mathrm{x1} = \mathrm{h1(y2)}$ and $\mathrm{x2} = \mathrm{h2(y2)}$ for the fixed $\mathrm{Y1} = \mathbf{y1} + \mathbf{\Delta y1}$.

For the curve 'c': The invertible equations for the equation $\mathrm{g2(x1, x2)} = \mathrm{y2}$ can be written as $\mathrm{x1} = \mathrm{h1(y1)}$ and $\mathrm{x2} = \mathrm{h2(y1)}$ for the fixed $\mathrm{Y2} = \mathbf{y2}$.

For the curve 'd': The invertible equations for the equation $\mathrm{g2(x1, x2)} = \mathrm{y2} + \Delta\mathrm{y2}$ can be written as $\mathrm{x1} = \mathrm{h1(y1)}$ and $\mathrm{x2} = \mathrm{h1(y1)}$ for the fixed $\mathrm{Y2} = \mathbf{y2} + \mathbf{\Delta y2}$.

Thus the rectangle with the co-ordinates mentioned and the corresponding parallelogram with the co-ordinates marked is as shown in the Fig. 2.14.

Point 1 (p1) on the curve 'b' can be obtained as x1+ (rate of change of h1 with respect to y2)(dy2) for the X1 co-ordinate and x2+ (rate of change of h2 with respect to y2)*dy2.

Point 2 (p2) on the curve 'c' can be obtained as x1+ (rate of change of h1 with respect to y1)(dy1) for the X1 co-ordinate and x2+ (rate of change of h2 with respect to y1)*dy1.

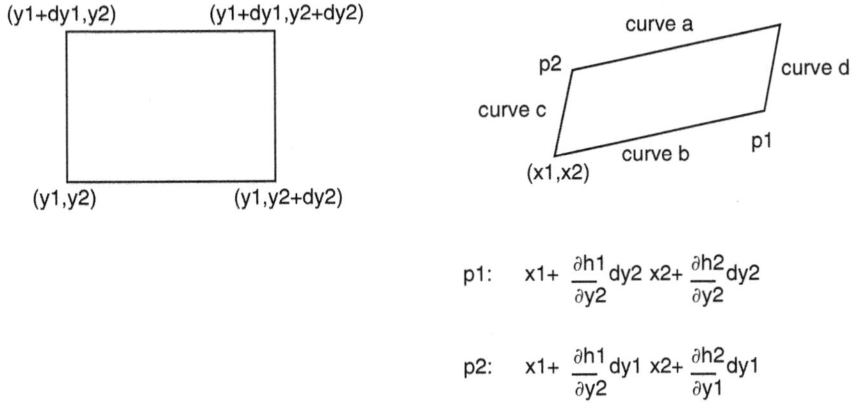

Fig. 2.14 Rectangle to Parallelogram mapping for the transformation of random variable

Thus the co-ordinates for the points ‘p1’ and ‘p2’ are obtained as follows.

$$\text{P1}: \left(x1+\frac{dh1}{dy2}dy2, x2+\frac{dh2}{dy2}dy2\right)$$
$$\text{P2}: \left(x1+\frac{dh1}{dy1}dy1, x2+\frac{dh2}{dy1}dy1\right)$$

Thus the equation becomes,

$$f_{Y1Y2}(y1, y2)\ [\text{Area of the rectangle}] = f_{X1X2}(x1, x2)\ [\text{Area of the Area of the parallelogram}]$$

Area of the Rectangle $= dy1\, dy2$

$$\text{Area of the parallelogram} = \begin{vmatrix} \frac{dh1}{dy2}dy2 & \frac{dh2}{dy2}dy2 \\ \frac{dk1}{dy1}dy1 & \frac{dh2}{dy1}dy1 \end{vmatrix} = \begin{vmatrix} \frac{dh1}{dy2} & \frac{dh2}{dy2} \\ \frac{dh1}{dy1} & \frac{dh2}{dy1} \end{vmatrix} dy1\, dy2$$

$$f_{Y1Y2}(Y1, y2)dy1\, dy2 - f_{X1X2}(x1, x2) \begin{vmatrix} \frac{dh1}{dy2} & \frac{dh2}{dy2} \\ \frac{dh1}{dy1} & \frac{dh2}{dy1} \end{vmatrix} dy1\, dy2$$

$$\Rightarrow f_{1\ 1Y2}(y1, y2) = f_{X1X2}\,(x1, x2) \begin{vmatrix} \frac{dh1}{dy2} & \frac{dh2}{dy2} \\ \frac{dh1}{dy1} & \frac{dh2}{dy1} \end{vmatrix}$$

The matrix $\begin{vmatrix} \frac{dh1}{dy2} & \frac{dh2}{dy2} \\ \frac{dh1}{dy1} & \frac{dh2}{dy1} \end{vmatrix}$ is called as Jacobian matrix.

Also it can be shown using the same procedure

$$f_{X1X2}(x1, x2) = f_{Y1Y2}(y1, y2) \begin{vmatrix} \frac{dg1}{dx2} & \frac{dg2}{dx2} \\ \frac{dg1}{dx1} & \frac{dg2}{dx1} \end{vmatrix}$$

Example 2.4. The random variables X and Y are related via Y = g(X), where g(.) is monotonically increasing function.

$$F_Y(y) = P(Y \le y) = P(\mathrm{g(X)} \le \mathrm{Y}) = \mathrm{P(X} \le \mathrm{g}^{-1}\mathrm{(y))} = F_X(\mathrm{g}^{-1}\mathrm{(y))}$$
$$\text{Let a} = \mathrm{g}^{-1}(\mathrm{y})$$
$$\Rightarrow \mathrm{y} = \mathrm{g(a)}$$
$$\Rightarrow F_Y(g(a)) = F_X(a)$$

Example 2.5. Let X be a random variable with uniform distribution over the interval [−4 to 4] (Fig. 2.15). Let the random variable $Y = g(X)$.

$$\begin{aligned} \textit{where } g(x) &= x, & |x| \le 2 \\ &= -2, & X < -2 \\ &= 2, & X > 2 \end{aligned}$$

The density function of the random variable Y is computed as follows.

$$P(Y \le y) = P(g(X)y) = P(X \le \mathrm{g}^{-1}(\mathrm{y}))$$

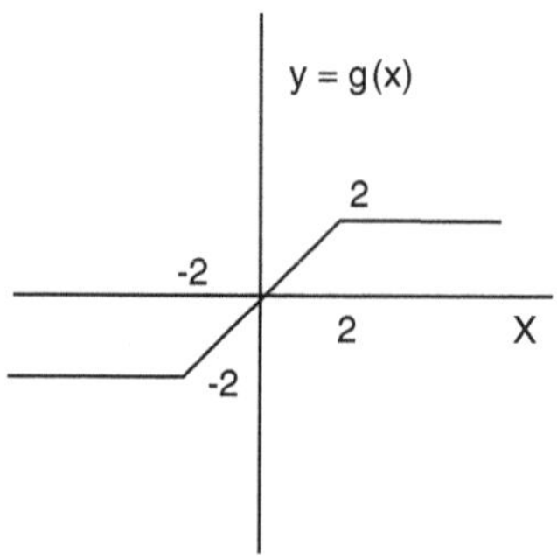

Fig. 2.15 Y = g(X) of the Example 2.5

From the graph (Fig. 2.16)

$$\begin{aligned} F_Y(\infty) &= P(Y \leq \infty) = P(X \leq \infty) = 1 \\ F_Y(2) &= P(Y \leq 2) = P(X \leq \infty) = 1 \\ F_Y(y) &= P(Y \leq y) = P(X \leq y) \, for \, -2 \leq y \leq 2 \\ \Rightarrow &= P(Y \leq -2) = 0 \end{aligned}$$

$f_Y(y)$ is obtained by differentiating $F_Y(y)$ with respect to y as shown in the graph below.

Note that there are two impulses in $f_Y(y)$. One at y = −2 and another at y = 2 ⇒ $P(Y = -2) = \frac{1}{4}$ *and* $P(Y = 2) = \frac{1}{4}$

Example 2.6. Let X and Y are the two random variables such that $Y = X^2$. Given probability density function of the random variable X, the probability density function of the random variable Y is obtained as follows (Fig. 2.17).

From the graph, $F_Y(y) = P(Y \leq y) = P(-\sqrt{y} \leq X \leq \sqrt{y}) = F_X(\sqrt{y}) - F_X(-\sqrt{y})$

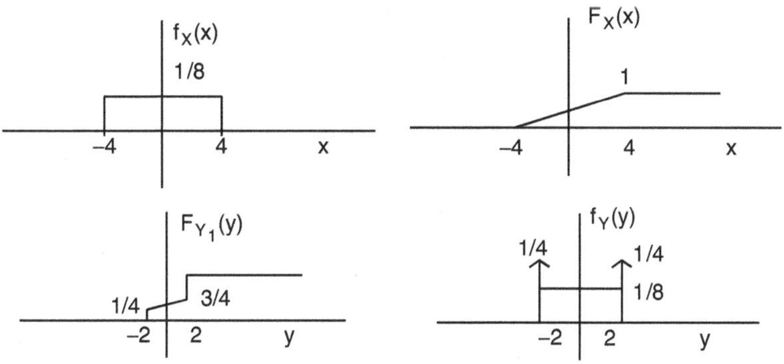

Fig. 2.16 PDF and the corresponding CDF of the random variables X and Y of the Example 2.5

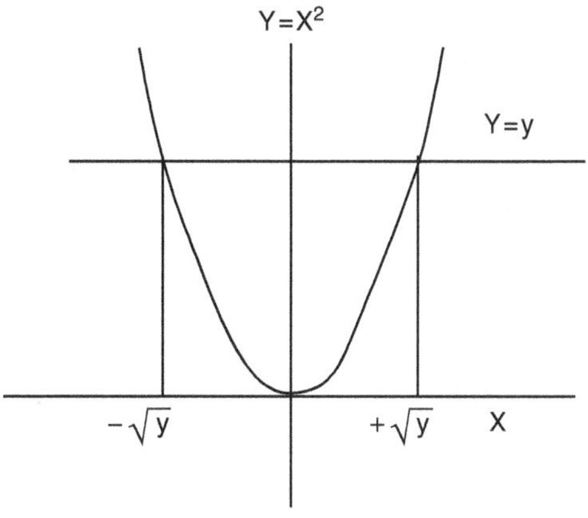

Fig. 2.17 Y = g(X) of the Example 2.6

Differentiating on both sides gives

$$f_Y(y) = \frac{1}{2\sqrt{y}}\left[f_X\left(\sqrt{y}\right) + f_X(-\sqrt{y})\right] \; for\ y \leq 0$$

(a) If X is having the following probability density function

$$\begin{aligned} f_X(x) &= \tfrac{x}{\alpha}e^{-\frac{x^2}{2\alpha}} x \geq 0 \\ &= 0,\ elsewhere \end{aligned}$$

$$\Rightarrow f_Y(y) = \frac{1}{2\sqrt{y}}\left[\frac{\sqrt{y}}{a}e^{-\frac{y}{2\alpha}} - 0\right]$$

[Note that $f_x(x) = 0\ for\ x < 0$]

$$\Rightarrow f_Y(y) = \left[\frac{1}{2\alpha}e^{-\frac{y}{2\alpha}}\right] \; for\ y \geq 0$$

(b) If X is having the following probability density function

$$f_X(x) = \left(\frac{1}{\sqrt{2\Pi\sigma^2}}e^{-\left(\frac{x^2}{2\sigma^2}\right)} for - \infty \leq x \leq \infty\right.$$

$$\Rightarrow f_Y(y) = \frac{1}{2\sqrt{y}}\left[\frac{1}{\sqrt{2\Pi\sigma^2}}e^{-\left(\frac{y}{2\sigma^2}\right)} + \frac{1}{\sqrt{2\Pi\sigma^2}}e^{-\left(\frac{y}{2\sigma^2}\right)}\right]$$

$$\Rightarrow f_Y(y) = \frac{1}{\sqrt{2\Pi y\sigma^2}}e^{-\left(\frac{y}{2\sigma^2}\right) for\ y \geq 0}$$

(c) $f_Y(y/X > 0)$ is computed as follows

$$\begin{aligned} F_Y(y/X > 0) &= P(y \leq Y/X > 0) \\ &= \frac{P(y \leq Y, X > 0)}{P(X > 0)} \\ &= \frac{P\left(-\sqrt{y} \leq X \leq \sqrt{y}, X > 0\right)}{P(X > 0)} \\ &= \frac{P\left(0 < X \leq \sqrt{y}\right)}{P(X > 0)} \\ &= \frac{F_X\left(\sqrt{y}\right) - F_X(0)}{P(X > 0)} \\ &= \frac{F_X\left(\sqrt{y}\right) - F_X(0)}{[1 - P(X \leq 0)} \\ &= \frac{F_X\left(\sqrt{y}\right) - F_X(0)}{[1 - F_X(0)]} \end{aligned}$$

Differentiating on both sides gives

$$\mathrm{f}_Y(y/X>0) = \left(\frac{1}{2\sqrt{y}}\right)\left[\frac{\mathrm{F}_X(\sqrt{y}))}{[1-\mathrm{F}_X(0)]}\right] \quad for\ y \geq 0$$

Example 2.7. Let X and Y are the two **independent** random variables, the probability density function of the random variable Z = X + Y is computed as follows (Fig. 2.18).

From the Graph

$$\mathrm{F}_Z(z) = P(Z \leq z) = P((X+Y) \leq z) = \int_{-\infty}^{\infty}\int_{-\infty}^{z-y} \mathrm{f}_{XY}(x,y)dx\,dy$$

$$\Rightarrow \mathrm{f}_Z(z) = \frac{\partial \int_{-\infty}^{\infty}\int_{-\infty}^{z-y} \mathrm{f}_{XY}(x,y)dx\,dy}{\partial z}$$

Using Leibnitz integration Formula

$$\Rightarrow \mathrm{f}_Z(z) = \int_{-\infty}^{\infty}\left[\frac{\partial(z-y)}{\partial z}\mathrm{f}_{XY}(z-y,y) - \frac{\partial(-\infty)}{\partial z}\mathrm{f}_{XY}(-\infty,y) + \int_{-\infty}^{z-y}\frac{\partial \mathrm{f}_{XY}(x,y)}{\partial z}dx\right]dy$$

$$\Rightarrow \mathrm{f}_Z(z) = \int_{-\infty}^{\infty} \mathrm{f}_{XY}(z-y,y)\,dy$$

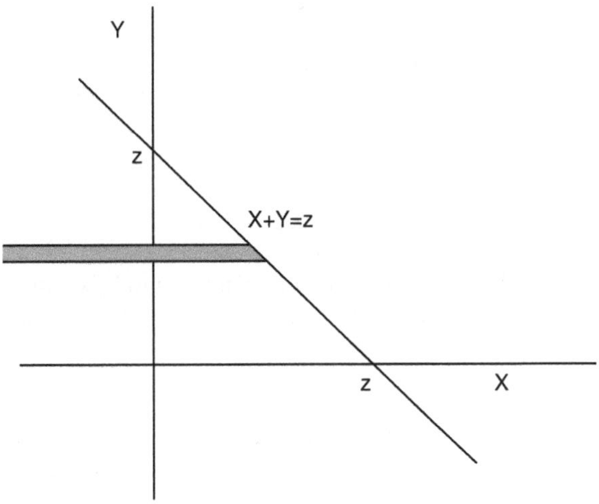

Fig. 2.18 Graph depicting X + Y = z

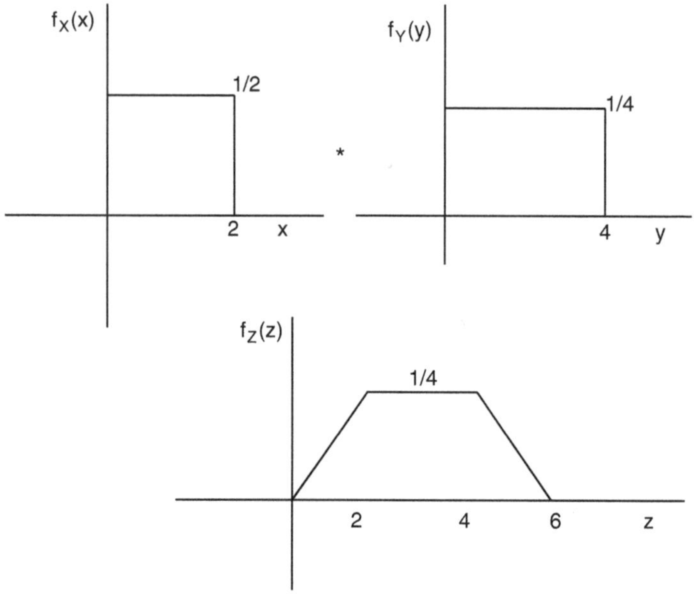

Fig. 2.19 Probability density function of the random variables X, Y and Z, where Z = X + Y

Note: Leibnitz integration
Differentiation of the integration with limits (Leibnitz integration)

$$g(x) = \int_{b(x)}^{a(x)} f(x,y)dy$$

$$\frac{dg(x)}{dx} = \frac{da(x)}{dx} f(x,a(x)) - \frac{db(x)}{dx} f(x,b(x) + \int_{b(x)}^{a(x)} \frac{\partial f(x,y)}{\partial x} dy$$

X and Y are **independent** $\Rightarrow f_Z(z) = \int_{-\infty}^{\infty} f_X(z-y) f_Y(y)\, dy$

$$\Rightarrow f_Z(z) = f_X(z) * f_Y(z)$$

Suppose if X is uniformly distributed between 0 and 2 and Y is uniformly distributed between 0 and 4, then the probability density function of Z is computed as shown below (Fig. 2.19).

Example 2.8. Let X and Y are the two **independent** random variables, the probability density function of the random variable Z = X/Y is computed as follows.

$$F_Z(z) = P(Z \le z) = P\left(\left(\frac{X}{Y}\right) \le z\right)$$

To obtain the solution the Trick is to Compute

$$
\begin{aligned}
\mathrm{F}_{\frac{Z}{Y}=y}(z) &= P\left(Z \le \frac{z}{Y} = y\right) \\
&= P(Z \le z, Y = y)/P(Y = y) \\
\Rightarrow \mathrm{F}_{Z/=y}(z) &= P\left((X/\left(Y = y\right)) \le z, Y = y\right)/P(Y = y) \\
\Rightarrow \mathrm{F}_{Z/Y=y}(z) &= P(X \le zy)P(Y = y)/P(Y = y) \\
&\text{[Because X and Y are independent]} \\
\Rightarrow \mathrm{F}_{Z/Y=y}(z/Y = y) &= P(X \le zy) = \mathrm{F}_X(zy)
\end{aligned}
$$

Differentiating on both sides gives,

$$\Rightarrow \mathrm{f}_{Z/Y=y}(z) = \mathrm{y}\,\mathrm{f}_X(zy)$$

Also

$$
\begin{aligned}
P(Z \le z/Y = y) &= \lim_{\Delta y \to 0} \frac{P(Z \le z, y \le Y \le y + \Delta y)}{P(y \le Y \le y + \Delta y)} \\
\Rightarrow F_{Z/Y=y}(x) &= \lim_{\Delta y \to 0} \frac{F_{ZY}(z, y + \Delta y) - F_{ZY}(z, y)}{F_Y(y + \Delta y) - F_Y(y)} \\
\Rightarrow F_{Z/Y=y}(x) &= \lim_{\Delta y \to 0} \frac{\left[F_{ZY}(z, y + \Delta y) - F_{ZY}(z, y)\right]/\Delta y}{\left[\mathrm{F}_Y(y + \Delta y) - F_Y(y)\right]/\Delta y} \\
\Rightarrow F_{Z/Y=y}(x) &= \frac{\frac{dF_{ZY}(z,\, y)}{dy}}{f_Y(y)} \\
\Rightarrow f_{Z/Y=y}(x) &= \frac{\frac{\partial^2 F_{ZY}(z,\, y)}{\partial z \partial y}}{f_Y(y)} \\
\Rightarrow f_{Z/Y=y}(x) &= \frac{f_{ZY}(z,\ y)}{f_Y(y)} \\
\Rightarrow \int f_{Z/Y=y}(x) f_Y(y)dy &= f_Z(z) \\
\Rightarrow f_Z(z) &= \int f_{Z/Y=y}(z) f_Y(y)dy \\
\mathrm{f}_{Z/Y=y}(z) &= \mathrm{y}\,\mathrm{f}_X(zy) \\
\Rightarrow f_Z(z) &= \int \mathrm{y}\,\mathrm{f}_X(zy)\mathrm{f}_Y(y)dy
\end{aligned}
$$

Example 2.9. The random variable X and Y are independent and identically distributed random variables. Let Z $= \sqrt{X^2 + Y^2}$ and $\Phi = tan^{-1}\left(\frac{Y}{X}\right)$. The joint density function $\mathrm{f}_{Z\Phi}(Z, \Phi)$ is computed as follows.

We know,

$$f_{Z\Phi}(z,\Phi) = \frac{f_{XY}(x,y)}{\begin{vmatrix} \frac{dg1}{dx} & \frac{dg2}{dx} \\ \frac{dg1}{dy} & \frac{dg2}{dy} \end{vmatrix}} at\ (x1,y1) + \frac{f_{XY}(x,y)}{\begin{vmatrix} \frac{dg1}{dx} & \frac{dg2}{dx} \\ \frac{dg1}{dy} & \frac{dg2}{dy} \end{vmatrix}}\ at\ (x2,y2)$$
$$+\cdots \frac{f_{XY}(x,y)}{\begin{vmatrix} \frac{dg1}{dx} & \frac{dg2}{dx} \\ \frac{dg1}{dy} & \frac{dg2}{dy} \end{vmatrix}}\ at\ (xn,yn)$$

Where $Z = \mathrm{g1}(\mathrm{X},\mathrm{Y})$ and $\Phi = \mathrm{g2}(\mathrm{X},\mathrm{Y})$ and (x1,y1), (x2,y2), … (xn,yn) are the solutions obtained by solving the equations g1(x,y) and g2(x,y) for the constant 'Z' and 'Φ'.

In our case, $\mathrm{g1}(\mathrm{X},\mathrm{Y}) = \mathrm{Z} = \sqrt{X^2+Y^2}$ and $\mathrm{g2}(\mathrm{X},\mathrm{Y}) = \Phi = tan^{-1}\left(\frac{Y}{X}\right)$. Solving for X and Y for the fixed Z and $\varPhi$ gives $\mathrm{X1} = \mathrm{Z}\cos\varPhi$ and $\mathrm{Y1} = \mathrm{Z}\sin\varPhi(\mathrm{x1},\ \mathrm{y1}) = (Zcos\varPhi,\ Zsin\varPhi)$

$$\frac{dg1}{dX} = \frac{X}{\sqrt{X^2+Y^2}} at\ (\mathrm{x1},\mathrm{y1}) = (Zcos\varPhi, Zsin\varPhi) = \frac{Zcos\varPhi}{Z} = cos\ \varPhi$$
$$\frac{dg1}{dY} = \frac{Y}{\sqrt{X^2+Y^2}} at\ (\mathrm{x1},\mathrm{y1}) = (Zcos\varPhi, Zsin\varPhi) = \frac{Zsin\varPhi}{Z} = sin\ \varPhi$$
$$\frac{dg2}{dX} = \frac{\left(-\frac{Y}{X^2}\right)}{1+(Y/X)^2} = \frac{-Y}{X^2+Y^2} at\ (\mathrm{x1},\mathrm{y1}) = (Zcos\varPhi, Zsin\varPhi)$$
$$= \frac{-ZSin\varPhi}{Z^2} = \frac{-Sin\varPhi}{Z}$$
$$\frac{dg2}{dY} = \frac{\left(\frac{1}{X}\right)}{1+(Y/X)^2} = \frac{X}{X^2+Y^2} at(\mathrm{x1},\mathrm{y1}) = (Zcos\varPhi, Zsin\varPhi) = \frac{cos\varPhi}{Z}$$
$$\Rightarrow \begin{vmatrix} \frac{dg1}{dx} & \frac{dg2}{dx} \\ \frac{dg1}{dy} & \frac{dg2}{dy} \end{vmatrix} = \begin{vmatrix} cos\varPhi & \frac{-Sin\varPhi}{Z} \\ sin\varPhi & \frac{cos\varPhi}{Z} \end{vmatrix} = 1/|\mathrm{Z}|$$

Thus $f_{Z\Phi}(z,\Phi) = \frac{f_{XY}(x,y)}{\left\|\begin{matrix} \frac{dg1}{dx} & \frac{dg2}{dx} \\ \frac{dg1}{dy} & \frac{dg2}{dy} \end{matrix}\right\|} at\ (x1,y1) = (Zcos\varPhi, Zsin\varPhi)$

$$\Rightarrow f_{Z\Phi}(z,\Phi) = |\mathrm{Z}|\, f_{XY}(Zcos\varPhi, Zsin\varPhi)$$

Suppose if $f_{XY}(x, y) = \frac{1}{2\Pi\sigma^2} e^{\frac{-(x^2+y^2)}{2\sigma^2}}$, where $-\infty \leq x, y \leq \infty$.

$$\Rightarrow f_{Z\Phi}(z, \Phi) = \frac{|z|}{2\Pi\sigma^2} e^{\frac{-(z^2)}{2\sigma^2}}, \ where\ Z \geq 0\ and\ 0 \leq \Phi \leq 2\Pi$$
$$= 0, \text{ Otherwise}$$

Example 2.10. The random variable X and Y are independent and identically distributed random variables. Let Z = $\sqrt{X^2 + Y^2}$ and W = X/Y. The joint density function

$f_{ZW}(Z, W)$ is computed as follows.

We know,

$$f_{ZW}(z, w) = \frac{f_{XY}(x, y)}{\begin{vmatrix} \frac{dg1}{dx} & \frac{dg2}{dx} \\ \frac{dg1}{dy} & \frac{dg2}{dy} \end{vmatrix}}\ at\ (x1, y1) + \frac{f_{XY}(x, y)}{\begin{vmatrix} \frac{dg1}{dx} & \frac{dg2}{dx} \\ \frac{dg1}{dy} & \frac{dg2}{dy} \end{vmatrix}}\ at\ (x2, y2)$$
$$+ \cdots \frac{f_{XY}(x, y)}{\begin{vmatrix} \frac{dg1}{dx} & \frac{dg2}{dx} \\ \frac{dg1}{dy} & \frac{dg2}{dy} \end{vmatrix}}\ at\ (xn, yn)$$

Where Z = g1(X, Y) and W = g2(X, Y) and (x1,y1), (x2,y2),...(xn,yn) are the solutions obtained by solving the equations g1(x,y) and g2(x,y) for the constant ‘Z’ and ‘W’.

In our case, g1(X, Y) = Z = $\sqrt{X^2 + Y^2}$ and g2(X, Y) = W = Y/X.

Solving for X and Y for the fixed Z and W gives the following set of solutions

$$Z = \sqrt{X^2 + Y^2} \Rightarrow Z^2 = X^2 + Y^2 = X^2 + (XW)^2 = X^2(1 + W^2)$$
$$\Rightarrow X^2 = Z^2/(1 + W^2)$$

$\Rightarrow$ X = +Z/$\sqrt{1 + W^2}$ or X = −Z/$\sqrt{1 + W^2}$ and the corresponding values for Y are

Y = +ZW/$\sqrt{1 + W^2}$ or Y = −ZW/$\sqrt{1 + W^2}$ respectively.

Therefore the solutions are $\left(Z/\sqrt{1 + W^2}, +ZW/\sqrt{1 + W^2}\right)$ and $(-Z/\sqrt{1 + W^2}, -ZW/\sqrt{1 + W^2})$

$$\frac{dg1}{dX} = \frac{X}{\sqrt{X^2 + Y^2}}\ at\ (x1, y1) = \left(+Z/\sqrt{1 + W^2}, +ZW/\sqrt{1 + W^2}\right)$$
$$= 1/\sqrt{1 + W^2}$$

$$\frac{dg1}{dX} = \frac{X}{\sqrt{X^2+Y^2}}\ at\ (\text{x2}, \text{y2}) = \left(-\frac{\text{Z}}{\sqrt{1+W^2}}, -\text{ZW}/\sqrt{1+W^2}\right)$$
$$= -\frac{1}{\sqrt{1+W^2}}$$
$$\frac{dg1}{dY} = \frac{Y}{\sqrt{X^2+Y^2}}\ at\ (\text{x1}, \text{y1}) = \left(+\text{Z}/\sqrt{1+W^2}, +\text{ZW}/\sqrt{1+W^2}\right)$$
$$= W/\sqrt{1+W^2}$$
$$\frac{dg1}{dY} = \frac{Y}{\sqrt{X^2+Y^2}}\ at\ (\text{x2}, \text{y2}) = \left(-\frac{\text{Z}}{\sqrt{1+W^2}}, -\text{ZW}/\sqrt{1+W^2}\right)$$
$$= -W/\sqrt{1+W^2}$$
$$\frac{dg2}{dX} = \frac{-Y}{X^2}\ at\ (\text{x1}, \text{y1}) = \left(+\text{Z}/\sqrt{1+W^2}, +\text{ZW}/\sqrt{1+W^2}\right)$$
$$= -\frac{W\sqrt{1+W^2}}{Z}$$
$$\frac{dg2}{dX} = \frac{-Y}{X^2}\ at\ (\text{x2}, \text{y2}) = \left(-\frac{Z}{\sqrt{1+W^2}}, -\text{ZW}/\sqrt{1+W^2}\right) = +\frac{W\sqrt{1+W^2}}{Z}$$
$$\frac{dg2}{dY} = \frac{1}{X}\ at\ (\text{x1}, \text{y1}) = \left(\frac{Z}{\sqrt{1+W^2}}, \text{ZW}/\sqrt{1+W^2}\right) = +\frac{\sqrt{1+W^2}}{Z}$$
$$\frac{dg2}{dY} = \frac{1}{X}\ at\ (\text{x2}, \text{y2}) = \left(-\frac{Z}{\sqrt{1+W^2}}, -\text{ZW}/\sqrt{1+W^2}\right) = -\frac{\sqrt{1+W^2}}{Z}$$
$$\text{Jacobian at (x1, y1)} = \begin{vmatrix} 1/\sqrt{1+W^2} & \frac{W\sqrt{1+W^2}}{Z} \\ -W/\sqrt{1+W^2} & \frac{\sqrt{1+W^2}}{Z} \end{vmatrix} = \frac{|(1+W^2)|}{|Z|}$$
$$\text{Jacobian at (x2, y2)} = \begin{vmatrix} -1/\sqrt{1+W^2} & -W/\sqrt{1+W^2} \\ \frac{W\sqrt{1+W^2}}{Z} & -\frac{\sqrt{1+W^2}}{Z} \end{vmatrix} = \frac{|(1+W^2)|}{|Z|}$$
$$f_{ZW}(z,w) = \frac{|Z|}{|(1+W^2)|}$$
$$\text{f}_{XY}\left(\frac{\text{Z}}{\sqrt{1+W^2}}, \text{ZW}/\sqrt{1+W^2}\right)$$
$$+\frac{|z|}{|(1+W^2)|}\text{f}_{XY}\left(-\frac{\text{Z}}{\sqrt{1+W^2}}, -\text{ZW}/\sqrt{1+W^2}\right)$$
$$\Rightarrow f_{ZW}(z,w) = \frac{|Z|}{|(1+W^2)|}\left(\text{f}_{XY}\left(\frac{Z}{\sqrt{1+W^2}}, ZW/\sqrt{1+W^2}\right)\right.$$
$$\left.+\text{f}_{XY}\left(\frac{-\text{Z}}{\sqrt{1+W^2}}, -\text{ZW}/\sqrt{1+W^2}\right)\right)$$
$$\left(\text{Z}/\sqrt{1+W^2}, +\text{ZW}/\sqrt{1+W^2}\right)$$

Suppose if $f_{XY}(x,y) = \frac{1}{2\Pi} e^{-(x^2+y^2)/2}$

$$\Rightarrow fzw(z,w) = \frac{|Z|}{|(1+W^2)|} \frac{1}{\Pi} e^{-((z)^2+(zw)^2)/2(1+W^2)}$$

$$\Rightarrow fzw(z,w) = \frac{|Z|}{|(1+W^2)|} \frac{1}{\Pi} e^{-\frac{z^2}{2}} \; for\; z \geq 0\, and\; -\infty \leq W \leq \infty$$

Given W = Y/X $-\infty \leq X \leq \infty$ *and* $-\infty \leq Y \leq \infty$ *and hence* $-\infty \leq W \leq \infty$

Example 2.11. Let X is a uniformly distributed random variable over the interval [0 1]. Y is the random variable related with the random variable X as Y = g(X). The invertible function 'g(.)' is related with the distribution function $F_Y(y)$ is as given below.

$$P(Y \leq y) = P(g(X) \leq y) = P(X \leq g^{-1}(y)) = F_X\left(g^{-1}(y)\right)$$

Note that g(.) must be the invertible function.

Also we know

$$F_X(x) = x \; as\; X\; is\; uniformly\; distributed\; over\; the\; interval\; [0\ 1]$$
$$\Rightarrow P(Y \leq y) = F_Y(y) = g^{-1}(y)$$
$$\Rightarrow F_Y(.) = g^{-1}(.)$$

For instance if $f_Y(y) = (e^{-\sqrt{2}|y|})/\sqrt{2}$, then g (.) is obtained as follows.
For $f_Y(y) = \frac{e^{-\sqrt{2}|y|}}{\sqrt{2}}$,
$F_Y(y)$ is computed as follows.

$$F_Y(y) = \int_{-\infty}^{y} \frac{e^{-\sqrt{2}|y|}}{\sqrt{2}} dy$$

Case 1: If $y \leq 0$

$$\int_{-\infty}^{y} \frac{e^{\sqrt{2}y}}{\sqrt{2}} dy$$
$$= \frac{e^{\sqrt{2}y}}{2}$$

Case 2: If $y \geq 0$

$$F_Y(y) = \int_{-\infty}^{0} \frac{e^{\sqrt{2}y}}{\sqrt{2}} dy + \int_{0}^{y} \frac{e^{-\sqrt{2}y}}{\sqrt{2}} dy$$

$$= \frac{1}{2} - \frac{e^{-\sqrt{2}y}}{2} + \frac{1}{2}$$

$$= 1 - \frac{e^{-\sqrt{2}}y}{2}$$

$$g^{-1}(y) = \frac{e^{\sqrt{2}y}}{2} = x \Rightarrow \sqrt{2}y/2 = \ln(x)$$

$$\Rightarrow y = \sqrt{2}\ln(x)\ for\ y \leq 0$$

$$\text{Similarly } for\ y \geq 0, g^{-1}(y) = 1 - \frac{e^{\sqrt{2}y}}{2} = x$$

$$\Rightarrow y = \ln(2(1-x))/\sqrt{2}\ for\ y \geq 0$$

$$\text{Thus the function } y = g(x) = \sqrt{2}\ln(x)\ for\ 0.5 \leq x \leq 1$$

$$= \ln(2(1-x))/\sqrt{2}\ for\ 0 \leq x \leq 0.5$$

Example 2.12. Let Z = max(X, Y) and W = min(X, Y), where, X and Y are arbitrary random variables. The joint density function $f_{ZW}(z, w)$ is computed in terms of $f_{XY}(x, y)$ is as follows.

When $z \geq w$, $P(Z \leq z, W \leq w) = P(X \leq z, Y \leq w) + P(X \leq w, Y \leq z) - P(X \leq w, Y \leq w)$ (See Fig. 2.20 given below)

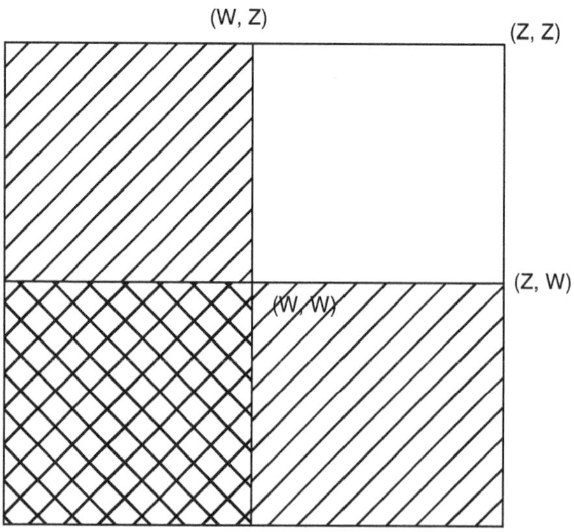

Fig. 2.20 Computation of the joint density function of the 2.12

$$\Rightarrow F_{ZW}(z,w) = F_{XY}(z,w) + F_{XY}(w,z) - F_{XY}(w,w) \ for \ z \geq w$$
$$= 0, \text{ Otherwise}$$
$$\Rightarrow f_{ZW}(z,w) = f_{XY}(z,w) + f_{XY}(w,z) - f_{XY}(w,w) \ for \ z \geq w$$
$$= 0, \text{ Otherwise}$$

For instance if X and Y are independent and identically distributed and is uniformly distributed between 0 and 4, then

$$f_{ZW}(z,w) = \frac{1}{25} for \ z \geq w, z, w = 0 \ to \ 4$$
$$= 0, \text{ Otherwise}$$

2.23 Expectations

Expectation of the random variable 'X' represented as E(X) is defined as follows:

$$\text{E(X)} = \int_{-\infty}^{\infty} x f_X(x)$$

Properties:

1. $\text{E(g(X))} = \int_{-\infty}^{\infty} g(x) f_X(x) dx$
2. $E_{X/Y=y} = \text{E(X/Y = y)} = \int_{-\infty}^{\infty} x f_{X/Y=y}(x)$
3. $\text{E(X)} = E_Y(E_{X/Y=y})$
4. If $X > 0$ then $E(X) > 0$
5. $\text{E}(X^2] = E[[X - E(X)]^2] + E[X]^2$

2.24 Indicator

(a) **Markov inequality** (Fig. 2.21)

$$\text{Consider} \quad X/b \geq \boldsymbol{I_b} \ \boldsymbol{(X)}$$
$$\Rightarrow E\left(\frac{X}{b}\right) \geq E\left(I_b(X)\right)$$
$$\Rightarrow E\left(\frac{X}{b}\right) \geq P(X \geq b)$$
$$\Rightarrow P(X \geq b) \leq \frac{E(X)}{b}$$

Fig. 2.21 Indicator

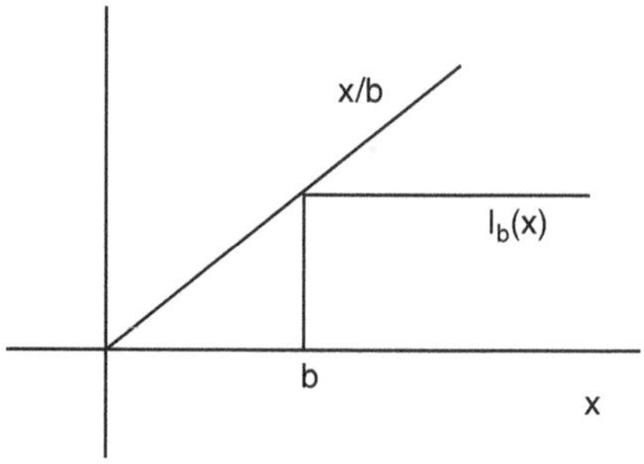

(b) **Chebyshev inequality**
Consider the random variable $X = Y^2$
Using Markov inequality

$$
\begin{aligned}
&P(Y^2 \geq b) \leq E(Y^2)/b^2 \\
&\Rightarrow P(|Y| \geq b) \leq E(Y^2)/b^2 \\
&\Rightarrow P(|Y - E(Y)| \geq b) \leq E(|Y - E(Y)|^2)/b^2 \\
&\Rightarrow P(|Y - E(Y)| \geq b) \leq \sigma_y^2/b^2
\end{aligned}
$$

(c) **Schwarz inequality**

$$
\begin{aligned}
&\text{Consider } E\left((aX - Y)^2\right) \geq 0 \\
&\Rightarrow a^2 E(X^2) + E(Y^2) - 2aE(XY) \geq 0 \\
&\Rightarrow a^2 E(X^2) - 2aE(XY) + E(Y^2) \geq 0
\end{aligned}
$$

The above equation can be viewed as the quadratic equation with variable 'a'
The equation is valid only when $4(E(XY))^2 - 4E(X^2)E(Y^2) \leq 0$

$$\Rightarrow (E(XY))^2 \leq E(X^2)E(Y^2)$$

(d) **Chernoff bound** (Fig. 2.22)

$$
\begin{aligned}
&\text{Consider } \boldsymbol{e^{SX-aS}} \geq \boldsymbol{I_a(x)} \\
&\Rightarrow E(\boldsymbol{e^{SX-aS}}) \geq \boldsymbol{E(I_a(x))} \\
&\Rightarrow E(\boldsymbol{e^{SX-aS}}) \geq P(X \geq b) \\
&\Rightarrow P(X \geq a) \leq \boldsymbol{e^{-aS}} E(\boldsymbol{e^{SX}})
\end{aligned}
$$

(e) Also it can be shown $E[XY] \leq 0.5(E(X^2) + E(Y^2))$ as follows

$$E((X - Y)^2) \geq 0$$

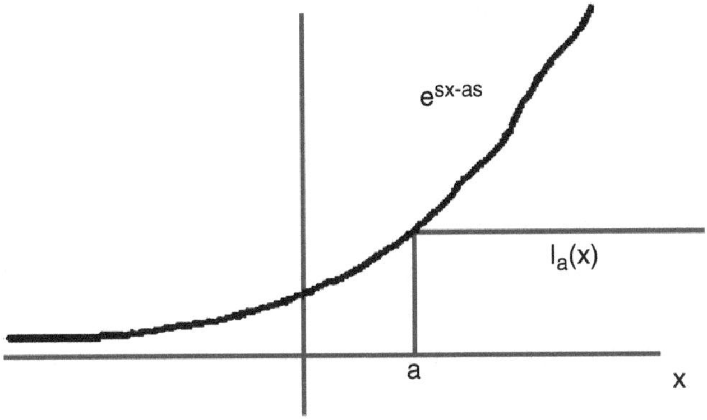

Fig. 2.22 Chernoff bound

$$\Rightarrow E(X^2) + E(Y^2) - 2E(XY)$$

$$\Rightarrow \text{E[XY]} \leq 0.5\left(E(X^2) + E(Y^2)\right)$$

(f) **Correlation co-efficient**
From Cauchy-Schwarz inequality $(E(XY))^2 \leq E(X^2)E(Y^2)$

$$\text{The ratio} \frac{(E(XY))^2}{E(X^2)E(Y^2)} \leq 1$$

$$\text{Define the ratio } \rho = \frac{E\left((X - m_X)(Y - m_Y)\right)}{\sqrt{E(X - m_X)^2)E((Y - m_Y)^2)}}$$

The ratio ρ is called correlation co-efficient. The range of ρ is given as $0 \leq |\rho| \leq 1$

2.25 Moment Generating Function

The moment generating function is defined as $\Phi_X(s) = \text{E}(e^{sX})$

Also it is known that $\text{E}(e^{sX}) = E\left(1 + \frac{(sX)^2}{2!} + \frac{(sX)^2}{3!} + \cdots\right.$

Therefore differentiating 'n-times' the moment generating function $\Phi_X(s)$ and equating s $= 0$ gives the $\text{E}(X^n)$.

Thus $\text{E}(X^n) = \frac{d^n \Phi_X(s)}{dS^n}|s = 0$

2.26 Characteristic Function

The characteristic function of the random variable X is given as $E(e^{jwX})$.
$E(X^n)$ can also be obtained using characteristic function.

2.27 Multiple Random Variable (Random Vectors)

Collections of random variables are called random vectors. The random vectors

represented as $\underline{X} = \begin{matrix} X1 \\ X2 \\ . \\ . \\ . \\ . \\ Xn \end{matrix}$

Joint Cumulative distribution function

$$F_{\underline{X}}(x1, x2, x3, \ldots xn) = pr(X1 \leq x1, X2, \leq x2, X3 \leq x3, \ldots Xn \leq xn)$$

Probability mass function of the random vector is represented as follows.

$$pr(X1 = x1, X2 = x2, X3 = x3, \ \ldots Xn = xn)$$

Probability density function of teh

$$f_{\underline{X}}(x1, x2, x3, \ldots xn) = \frac{\partial^n}{\partial x1 \partial x2 \partial x3 \ldots \partial xn} F_{\underline{X}}\ (x1, x2, x3, \ldots xn)$$

Similarly

$$F_{\underline{X}}(x1, x2, x3, \ldots xn) = \int\int\int \cdots \int f_{\underline{X}}\ (x1, x2, x3, \ldots xn)\ \partial x1 \partial x2 \partial x3 \ldots \partial xn$$

Marginal density function

$$f_{Xi}(xi) = \int\int\int \cdots \int_{\partial x1 \partial x2 \ldots \partial x(i-1) \partial x(i+1), \ldots \partial xn}^{f_{\underline{X}}\ (x1, x2, \ldots x(i-1), x(i+1) x(i) \ldots xn)}$$

Joint density function

$$f_{X1X2}(x1,x2) = \int\int\int\cdots\int f_{\underline{X}}\begin{pmatrix} x1,x2,\ldots x(i-1) \\ x(i),x(i+1)\ldots xn \end{pmatrix} \partial x3\partial x4\ldots\partial xn$$

Conditional probability density function

Consider the random vector $\underline{X}$ is divided into two random vectors $\underline{X1}$ and $\underline{X2}$ as shown below.

$$\underline{X} = \begin{bmatrix} \underline{X1} \\ \underline{X2} \end{bmatrix}$$

Then the conditional probability of the random vector $\underline{X1}$ over the random vector $\underline{X2}$ is computed as follows

$$\mathrm{f}_{\underline{X1}/\underline{X2}}(\mathrm{x1}) = \mathrm{f}_{\underline{X}}(\underline{x})/\mathrm{f}_{\underline{X2}}(\mathrm{x2})$$

Also note that

$$\mathrm{f}_{\underline{X}}(\underline{x}) = \mathrm{f}_{\underline{X1}/\underline{X2}}(\mathrm{x1})\mathrm{f}_{\underline{X2}}(\mathrm{x2})$$

In General,

$$\mathrm{f}_{\underline{X}}(\underline{x}) = f_{X1}(x1)f_{X2/X1}(x2)f_{X3/X1,X2}(x3)\ldots f_{Xn/Xn-1\ldots X1}(xn)$$

Independence

The random variables of the random vectors X1,X2,X3,. . .Xn are independent if

$$F_{\underline{X}}\ (x1,x2,x3,\ldots xn) = F_{X1}(x1)F_{X2}(x2)F_{X3}(x3)\ldots F_{Xn}(xn)$$
$$f_{\underline{X}}\ (x1,x2,x3,\ldots xn) = f_{X1}(x1)f_{X2}(x2)f_{X3}(x3)\ldots f_{Xn}(xn)$$

Expectation of the random vector

$$\mathrm{E}(\underline{X}) = \begin{matrix} E(X1) \\ E(X2) \\ . \\ . \\ . \\ . \\ E(Xn) \end{matrix}$$

Moment generating function of the random vector

$$\mathrm{E}(e^{\underline{S}^T\underline{X}})$$

Where

$$\underline{X} = \begin{matrix} X1 \\ X2 \\ . \\ . \\ . \\ . \\ Xn \end{matrix} \qquad \underline{S} = \begin{matrix} s1 \\ s2 \\ . \\ . \\ . \\ . \\ sn \end{matrix}$$

Characteristic function of the random vector

$$\mathrm{E}(e^{\underline{jw}^T \underline{X}})$$

where

$$\underline{X} = \begin{matrix} X1 \\ X2 \\ . \\ . \\ . \\ . \\ Xn \end{matrix} \qquad \underline{W} = \begin{matrix} w1 \\ w2 \\ . \\ . \\ . \\ . \\ wn \end{matrix}$$

Correlation matrix of the random vector

$$\mathrm{E}(\underline{X}\ \underline{X}^T) =$$
$$\begin{matrix} E(X1^2) & E(X1X2) & E(X1X3) & \ldots & E(X1Xn) \\ E(X2X1) & E(X2^2) & E(X2X3) & \ldots & E(X2Xn) \\ . & . & . & \ldots & . \\ . & . & . & \ldots & . \\ . & . & . & \ldots & . \\ E(XnX1) & E(XnX2) & E(XnX3) & \ldots & E(Xn^2) \end{matrix}$$

Covariance matrix of the random vector

$$\mathrm{E}\left((\underline{X} - E(\underline{X})(\underline{X} - E(\underline{X})^T\right)$$
$$= \mathrm{E}(\underline{X}\ \underline{X}^T) - E(X)E(X)^T$$

Note:

1. The events Xi and Xj are statistically uncorrelated if E(Xi Xj)−E(Xi)E(Xj) = 0, i ≠ j. Also note that the co-variance matrix becomes diagonal matrix if the elements of the random vector are uncorrelated to each other.
2. The events Xi and Xj are independent then E(XiXj) = E(Xi)E(Xj).
3. If the two events are statistically independent, they are uncorrelated. But the vice-versa is not true.

Gaussian probability density function with mean 'μ' and variance 'σ^2'

$$fx(x) = \frac{1}{\sqrt{2\Pi\sigma^2}}e^{-\frac{[x-\mu]^2}{2\sigma^2}}$$
$$F_X(x) = P(X \leq x)$$

Suppose $y^2 = \frac{[X-\mu]^2}{\sigma^2}$

$$\begin{aligned}
F_X(x) &= \int_{-\infty}^{x} \frac{1}{\sqrt{2\Pi\sigma^2}}e^{-\frac{[x-\mu]^2}{2\sigma^2}}dx \\
&= \int_{-\infty}^{\frac{[x-\mu]}{\sigma}} \frac{\sigma}{\sqrt{2\Pi\sigma^2}}e^{-\frac{y^2}{2}}dy \\
&= \int_{-\infty}^{\frac{[x-\mu]}{\sigma}} \frac{\sigma}{\sqrt{2\Pi\sigma^2}}e^{-\frac{y^2}{2}}dy \\
&= \int_{-\infty}^{\frac{[x-\mu]}{\sigma}} \frac{1}{\sqrt{2\Pi}}e^{-\frac{y^2}{2}}dy \\
&= F_Y\left(\frac{[X-\mu]}{\sigma}\right)
\end{aligned}$$

Thus Gaussian distribution value at 'x' with mean $= \mu$, variance $= \sigma^2$ can be computed using Gaussian distribution function of y at $\frac{[X-\mu]}{\sigma}$ whose mean $= 0$ and variance $= \sigma^2$

Moment generating function of the Gaussian density function

$$\Phi_X(s) = \mathrm{E}(e^{sx}] = \int_{-\infty}^{\infty} \frac{e^{sx}}{\sqrt{2\Pi\sigma^2}}e^{-\frac{[x-\mu]^2}{2\sigma^2}}dx$$

Consider the powers of e.

$$\begin{aligned}
&\frac{sx2\sigma^2 - x^2 - \mu^2 + 2\mu x}{2\sigma^2} \\
&= -\left(\frac{x^2 + \mu^2 - (2\mu + 2s\sigma^2)x}{2\sigma^2}\right)
\end{aligned}$$

Adding $(\mu + s\sigma^2)^2$ and subtracting $(\mu + s\sigma^2)^2$ on the numerator, we get the following

$$\begin{aligned}
&= -\left(\frac{x^2 + \mu^2 - (2\mu + 2s\sigma^2)x + (\mu + s\sigma^2)^2 - (\mu + s\sigma^2)^2}{2\sigma^2}\right) \\
&= -\left(\frac{(x - (2\mu + 2s\sigma^2))^2 + \mu^2 - (\mu + s\sigma^2)^2}{2\sigma^2}\right)
\end{aligned}$$

$$\Rightarrow \int_{-\infty}^{\infty} \frac{e^{-\frac{\mu^2-(\mu+s\sigma^2)^2}{2\sigma^2}}}{\sqrt{2\Pi\sigma^2}} e^{\frac{(x-(2\mu+2s\sigma^2))^2}{2\sigma^2}} dx$$

$$= e^{\mu s} e^{\frac{\sigma^2 s^2}{2}}$$

$$Because, \int_{-\infty}^{\infty} \frac{e^{-\frac{(x-(2\mu+2s\sigma^2))^2}{2\sigma^2}}}{\sqrt{2\Pi\sigma^2}} dx = 1$$

$$E[X] = \frac{\partial \Phi_X(s)}{\partial s} | s = 0$$

$$\mathrm{E}[X^n] = \frac{\partial^n \Phi_X(s)}{\partial S^n} | s = 0$$

$$\Phi_X(s) = e^{\mu s} e^{\frac{\sigma^2 s^2}{2}}$$

Consider the case when $\mu = 0$.

$$\Phi_X(s) = e^{\frac{\sigma^2 s^2}{2}}$$

$$= 1 + \frac{\frac{\sigma^2 s^2}{2}}{1!} + \frac{\left(\frac{\sigma^2 s^2}{2}\right)^2}{2!} + \ldots. + \frac{\left(\frac{\sigma^2 s^2}{2}\right)^{2m}}{2m!} + \cdots$$

$$\Rightarrow \frac{\partial^{2m} \Phi_X(s)}{\partial S^{2m}} | s = 0 = \frac{\sigma^{2m}}{2^m m!} (2m)(2m-1)\ldots 1$$

$$\text{Therefore } \mathrm{E}[X^n] = \sigma^n[1.3.5.7\ldots.(n-1)] \text{ when n is even}$$

$$= 0 \text{ when n is odd}$$

Chernoff bound for Gaussian density function with zero mean and unity variance

The Chernoff bound is given as $P(X \geq a) \leq e^{-aS} E(e^{SX})$ (i.e) $P(X \geq a) \leq e^{-aS} \Phi_X(s)$

For Gaussian density function $\Phi_X(s) = e^{\mu s} e^{\frac{\sigma^2 s^2}{2}}$

$$\Rightarrow P(X \geq a) \leq e^{-aS} e^{\mu s} e^{\frac{\sigma^2 s^2}{2}} \; for\ all\ s$$

To get the tight bound we have to find the value of 's' so that $e^{-aS} e^{\mu s} e^{\frac{\sigma^2 s^2}{2}}$ is minimized.

Assume $\mu = 0$ and variance $\sigma^2 = 1$ for simplicity.

Differentiating $e^{-aS} e^{\frac{\sigma^2 s^2}{2}}$ with respect to s and equate to zero

$$e^{-aS} e^{\frac{\sigma^2 s^2}{2}} \left(\frac{\sigma^2}{2}(2s)\right) + e\frac{\sigma^2 s^2}{2} e^{-aS}(-a) = 0$$

$$s = (a/\sigma^2) = a$$

Substituting s = a in $e^{-aS} e^{\frac{s^2}{2}}$ we get, $e^{\frac{-a^2}{2}}$

$$\Rightarrow P(X \geq a) \leq e^{\frac{-a^2}{2}}$$

2.28 Gaussian Random Vector with Mean Vector $\underline{\mu_X}$ and Covariance Matrix C_X

$$f_{\underline{X}}(\underline{x}) = \frac{1}{(2\Pi)^{\frac{n}{2}} |C|^{\frac{1}{2}}} e^{-} \left(\frac{1}{2}\right) \left(\left[\underline{X} - \underline{\mu_X}\right]^T C^{-1} \left[\underline{X} - \underline{\mu_X}\right] \right)$$

Moment Generating function for the Gaussian random variable is given as

$$\Phi_{\underline{X}}(\underline{S}) = e^{[\underline{\mu_X}]^T S} e^{\frac{1}{2}[S^T C S]}$$

Properties of Gaussian random vector

1. $E(\underline{X}) = \begin{matrix} E(X1) \\ E(X2) \\ \cdot \\ \cdot \\ \cdot \\ \cdot \\ E(Xn) \end{matrix} = \underline{\mu_X}$
2. $E(X_i) = \left.\frac{d\Phi_{\underline{X}}(\underline{s})}{dSi}\right| \underline{s} = 0$
3. $E\left(X_i{}^2\right) = \left.\frac{d^2\Phi_{\underline{X}}(s)}{dSi^2}\right| \underline{s} = 0$
4. $E(X_i X_j) = \left.\frac{d^2\Phi_{\underline{X}}(s)}{dS_i S_j}\right| \underline{s} = 0$
5. In general $E\left(X_i{}^m X_j{}^n\right) = \left.\frac{d^{m+n}\Phi_{\underline{X}}(s)}{dS_i{}^m s_j{}^n}\right| \underline{s} = 0$
6. The correlation matrix of the Gaussian random vector is given as

$$R = E(\underline{X}\,\underline{X}^T) = \begin{matrix} E(X1^2) & E(X1X2) & E(X1X3) & \ldots & E(X1Xn) \\ E(X2X1) & E(X2^2) & E(X2X3) & \ldots & E(X2Xn) \\ \cdot & \cdot & \cdot & \ldots & \cdot \\ \cdot & \cdot & \cdot & \ldots & \cdot \\ \cdot & \cdot & \cdot & \ldots & \cdot \\ E(XnX1) & E(XnX2) & E(XnX3) & \ldots & E(Xn^2) \end{matrix}$$

7. The co-variance matrix is obtained as

$$
\begin{aligned}
\mathrm{C} &= \mathrm{E}\left((\underline{X} - E(\underline{X})\,(\underline{X} - E(\underline{X})^T\right) \\
&= \mathrm{E}(\underline{X}\,\underline{X}^T) - E(\underline{X})E(\underline{X})^T
\end{aligned}
$$

8. Marginal density of any ℓ-dimensional sub vector of $\underline{X}(\ell \leq n)$ is also Gaussian with proper mean and co-variance matrix as described below

Consider the Gaussian random vector $\underline{X} = \begin{matrix} X_1 \\ X_2 \\ . \\ . \\ . \\ X_l \\ X_{l+1} \\ X_{l+2} \\ X_{l+3} \\ . \\ . \\ . \\ Xn \end{matrix}$

With mean vector $\underline{m} = \begin{matrix} m_1 \\ m_2 \\ . \\ . \\ . \\ m_l \\ m_{l+1} \\ m_{l+2} \\ m_{l+3} \\ . \\ . \\ . \\ m_n \end{matrix}$

And the covariance matrix given as

$$
\begin{matrix}
c_{11} & c_{12} & \ldots & c_{1l} & \ldots & c_{1n} \\
c_{21} & c_{22} & \ldots & c_{2l} & \ldots & c_{2n} \\
. & . & \ldots & . & \ldots & \\
. & . & \ldots & . & \ldots & \\
. & . & \ldots & . & \ldots & \\
c_{l1} & c_{l2} & \ldots & c_{ll} & \ldots & c_{ln}
\end{matrix}
$$

$$\begin{matrix} c_{l+1,1} & c_{l+1,2} & \cdots & c_{l+1,l} & \cdots & c_{l+1,n} \\ . & . & \cdots & . & \cdots & . \\ . & . & \cdots & . & \cdots & . \\ . & . & \cdots & . & \cdots & . \\ c_{n1} & c_{n2} & \cdots & c_{n,l} & \cdots & c_{nn} \end{matrix}$$

Let the sub vector of the above mentioned Gaussian vector be

$$\underline{Y} = \begin{matrix} X_1 \\ X_2 \\ . \\ . \\ . \\ X_l \end{matrix}$$

The random vector $\underline{Y}$ is also Gaussian distributed with mean vector

$$\underline{my} = \begin{matrix} m_1 \\ m_2 \\ . \\ . \\ . \\ m_l \end{matrix}$$

and co-variance matrix

$$\begin{matrix} c_{11} & c_{12} & . & . & . & c_{1l} \\ c_{21} & c_{22} & . & . & . & c_{2l} \\ . & . & . & . & . & . \\ . & . & . & . & . & . \\ . & . & . & . & . & . \\ c_{1l} & c_{l2} & . & . & . & c_{ll} \end{matrix}$$

9. If the Gaussian random vector $\underline{X}$ with mean vector $\underline{M}$ and co-variance matrix 'C_X' is linearly transformed into the random vector $\underline{Y}$ using the transformation matrix 'A' and the column vector $\underline{b}$ as

$$\underline{Y} = A\underline{X} + \underline{b}$$

Then the random vector $\underline{Y}$ is also Gaussian distributed with mean vector A $\underline{M} + \underline{b}$ and the co-variance matrix A $C_X A^T$
Note that the co-variance matrix of the random variable $\underline{Y}$ is A $C_X A^T$ irrespective of the distribution function. Similarly the mean vector of the random variable $\underline{Y}$ is A $\underline{M} + \underline{b}$ irrespective of the type of the distribution function.

10. Consider the random vector $\underline{X}$ is represented as row concatenation of two random vectors $\underline{X1}$ and $\underline{X2}$ as shown below

$$\underline{X} = \begin{bmatrix} \underline{X1} \\ \underline{X2} \end{bmatrix}$$

Similarly the mean vector the random vector $\underline{X}$ is represented as row concatenation of the mean vectors of the random vectors $\underline{X1}$ and $\underline{X2}$ as shown below.

$$\underline{M} = \begin{bmatrix} \underline{M1} \\ \underline{M2} \end{bmatrix}$$

Also let the co-variance matrix of the random vector $\underline{X1}$ and $\underline{X2}$ are represented as C_{11} and C_{22} respectively. The co-variance matrix of the random vector $\underline{X}$ is represented as

$$\begin{bmatrix} [C_{11}] & [C_{12}] \\ [C_{21}] & [C_{22}] \end{bmatrix}$$

(a) The conditional density function $f_{X2/X1=x1}(x2)$ is also Gaussian with mean vector $\underline{M2} + [C_{21}][C_{11}]^{-1}[\underline{X1} - \underline{M1}]$ and co-variance matrix $C = [C_{22}] - [C_{21}][C_{11}]^{-1}[C_{12}]$
(b) The conditional density function $f_{X1/X2=x2}(x1)$ is also Gaussian with mean vector $\underline{M1} + [C_{12}][C_{22}]^{-1}[\underline{X2} - \underline{M2}]$ and co-variance matrix $C = [C_{11}] - [C_{12}][C_{22}]^{-1}[C_{21}]$

11. If the co-variance matrix of the Gaussian random vector is diagonal, the elements in the random vector $\underline{X}$ are uncorrelated and independent
12. Contours of 2D-Gaussian probability density function (Fig. 2.23)

$$f_{XY}(x,y) = \frac{1}{2\Pi\sigma_1\sigma_2\sqrt{1-\rho^2}} e^{\frac{-\left(\frac{x^2}{\sigma_1{}^2} + \frac{y^2}{\sigma_2{}^2} - \frac{2\rho xy}{\sigma_1\sigma_2}\right)}{2(1-\rho^2)}}$$

Consider the 2D contour obtained from the equation

$$\frac{x^2}{\sigma_1{}^2} + \frac{y^2}{\sigma_2{}^2} - \frac{2\rho xy}{\sigma_1\sigma_2} = \text{constant} = \text{c1}$$

This is the contour having the same probability density value.

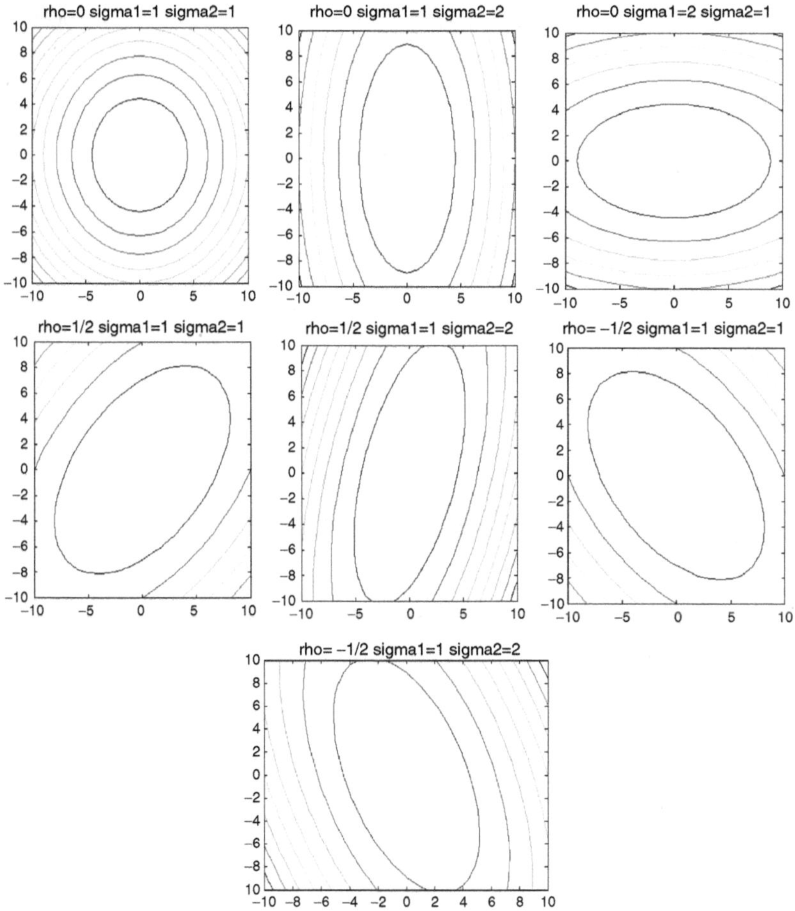

Case 1: $\rho = 0, \sigma_1 = \sigma_2$

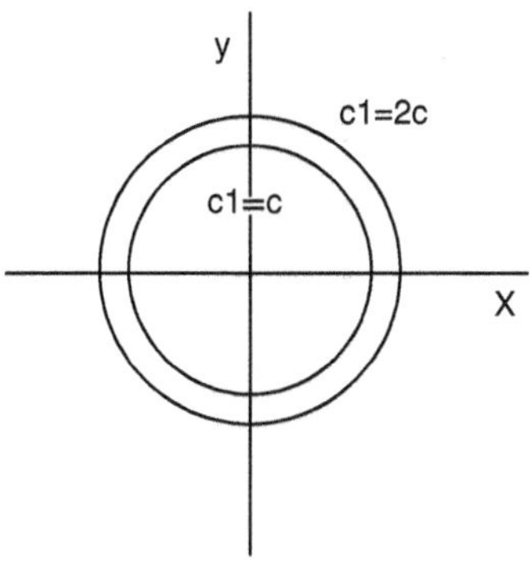

Fig. 2.23 Contours of the 2D Gaussian probability density function

Note that when the contour radius increases, actual magnitude of the pdf decreases. Thus contour c1 is having higher magnitude compared with the contour c2.
Case 2: $\rho = 0, \sigma_1 = \sigma_2$

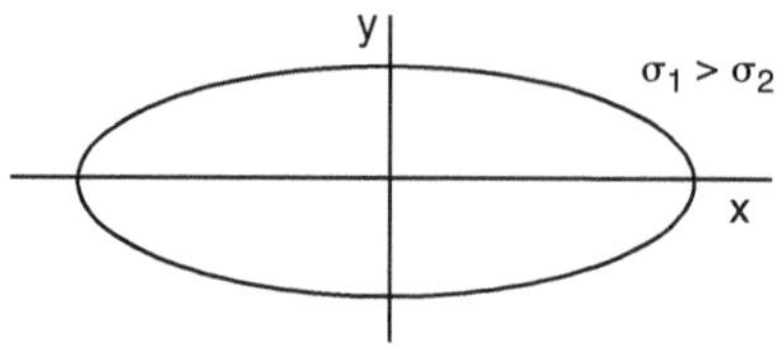

Case 3: $\rho = 0, \sigma_2 = \sigma_1$

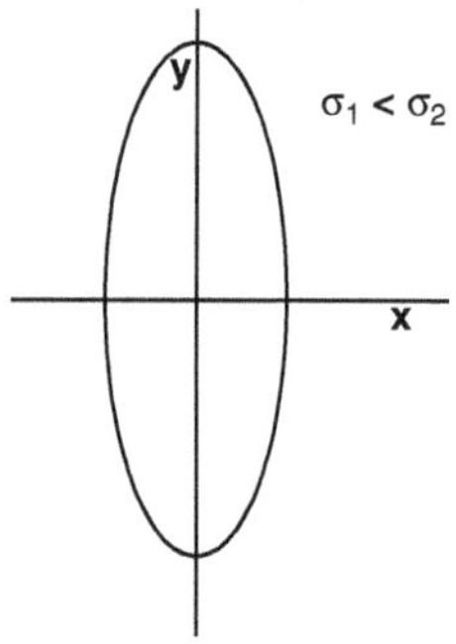

Case 4: $\rho < 0, \sigma_2 = \sigma_1$. Note that the value of $\Theta = 45°$

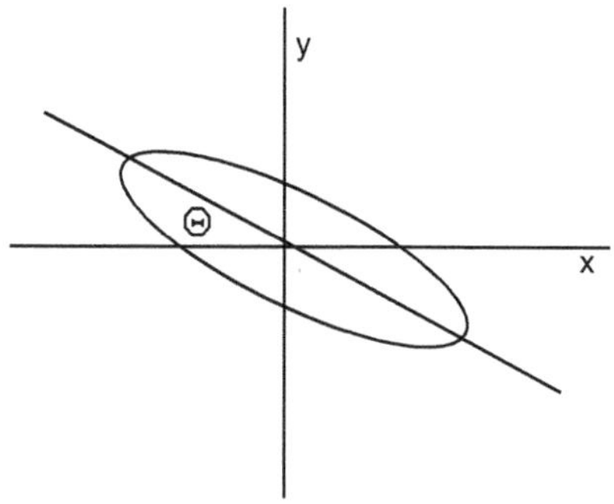

Fig. 2.23 (continued)

Case 5: $\rho > 0, \sigma_2 = \sigma_1$. Note that the value of $\Theta = 45°$

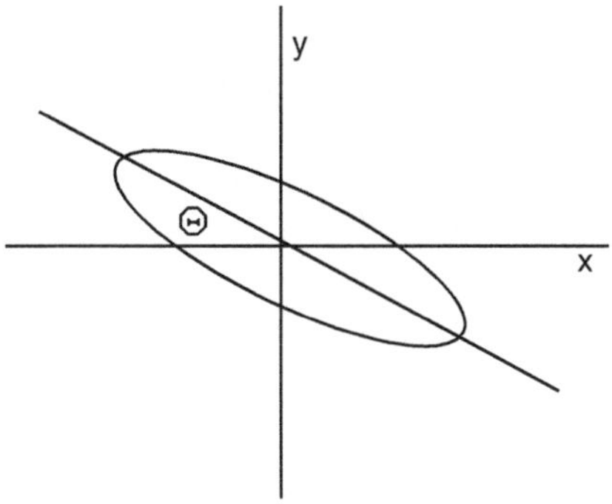

Case 6: $\rho \neq 0, \sigma_2 \neq \sigma_1$, Note that the value of $\Theta \neq 45°$

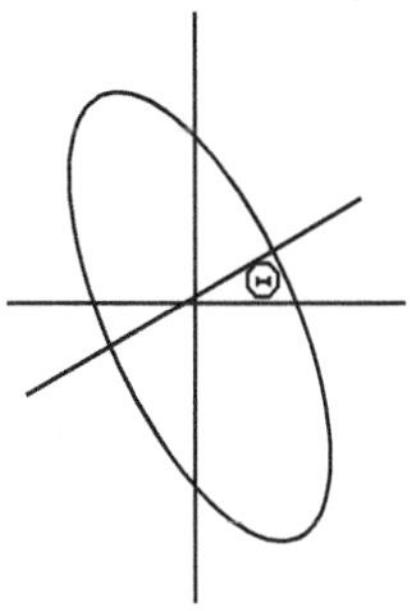

Fig. 2.23 (continued)

Example 2.13. Let X be a uniform random variable in [0,100]. $E(X/X \geq 65)$ is computed as follows.

$$E(X/X \geq 65) = \int_0^{100} x\ f_{X/X\geq 65}(x)dx$$

$f_{X/X\geq 65}$ is obtained as follows

$$\begin{aligned} F_{X/X\geq 65}(x) &= P(X < x, X \geq 65)/P(X \geq 65) \\ &= P(65 \leq X \leq x)/P(X \geq 65) \\ &= (F_X(x) - F_X(65))/P(X \geq 65) \\ &= (F_X(x) - F_X(65))/P(X \geq 65) \end{aligned}$$

Differentiating on both sides,

$$\Rightarrow f_{X/X\geq 65}(x) = f_X(x)/P(X \geq 65)\ for\ X \geq 65$$

$$\Rightarrow f_{X/X\geq 65}(x) = f_X(x) \Big/ \int_{65}^{100} f_X(x)dx$$

$$\Rightarrow f_{X/X\geq 65}(x) = f_X(x) \Big/ \int_{65}^{100} x\ f_X(d)dx$$

$$f_X(d) = \frac{1}{100}\ for\ 0 \leq x \leq 100 \text{ (Uniformly distributed)}$$

$$\Rightarrow f_{X/X\geq 65}(x) = f_X(x) \Big/ \int_{65}^{100} (1/100)dx$$

$$= (1/100) \Big/ \left(\frac{1}{100}\right)(35)$$

$$= \frac{1}{35}\ for\ X \geq 65\ and\ X \leq 100$$

$$\Rightarrow E(X/X \geq 65) = \int_{65}^{100} x\ f_{X/X\geq 65}(x)dx$$

$$= \int_{65}^{100} x\left(\frac{1}{35}\right)dx$$

$$= \left(\frac{1}{35}\right)\int_{65}^{100} x\ dx$$

$$= \frac{\left(\frac{1}{35}\right)(100^2 - 65^2)}{2} = \frac{165 * 35}{35 * 2}$$

$$= 82.5$$

Example 2.14. Let X be a poison random variable with probability mass function

$$P_X(k) = \frac{e^{-\lambda}\lambda^k}{k!}\ for\ k = 0, 1, 2, \ldots$$

E(X) and variance(X) are computed as follows

The moment generating function

$$\Phi_X(s) = E(e^{SX}) = \sum_{x=0}^{\infty} e^{Sx}\frac{e^{-\lambda}\lambda^x}{x!}$$

$$= \sum_{x=0}^{\infty} \frac{(e^S\lambda)^x e^{-\lambda}}{x!}$$

$$= e^{-\lambda}\left[1 + \frac{(e^S\lambda)}{1!} + \frac{(e^S\lambda)^2}{2!} + \frac{(e^S\lambda)^3}{3!} + \cdots\right]$$

$$= e^{-\lambda}e^{(e^S\lambda)}$$

Differentiating on both sides with respect to 's' and substitute s = 0 gives

$$\mathrm{E(X)} = e^{-\lambda}\frac{de^{(e^S\lambda)}}{ds}(at\ s=0) = e^{-\lambda}e^{(e^S\lambda)}\lambda e^S\ (at\ s=0) = \lambda$$

Differentiating $e^{-\lambda}\ e^{(e^S\lambda)}\lambda e^S$ with respect to 's' and substitute s = 0 gives

$$E(X^2) = \lambda e^{-\lambda}\left[e^S e^{(e^S\lambda)}\lambda e^S + e^{(e^S\lambda)}e^S\right](at\ s=0) = (\lambda^2+\lambda)$$

$$\Rightarrow \text{Variance} = [E(X^2) - \mathrm{E(X)}^2] = \lambda$$

$$\text{Mean} = \lambda$$

$$\text{Variance} = \lambda$$

Example 2.15. Consider the random variable X with probability density function as given below.

$$f_X(x) = \lambda e^{-\lambda x}\ for\ x \geq 0$$
$$= 0,\ \text{otherwise}$$

E(X), $f_X(x/X \geq 2)$ and $E(X/X \geq 2)$ are computed as shown below.

$$\mathrm{E(X)} = \int_{-\infty}^{\infty} x\ f_X(x)dx = \int_0^{\infty} x\ \lambda e^{-\lambda x}dx = \lambda\int_0^{\infty} x\ e^{-\lambda x}dx$$

$$= \lambda\int_0^{\infty} x\ d\left[\frac{e^{-\lambda x}}{-\lambda}\right]$$

$$== \lambda\left[x\left[\frac{e^{-\lambda x}}{-\lambda}\right]_0^{\infty} - \int_0^{\infty}\frac{e^{-\lambda x}}{-\lambda}dx\right] = \lambda[1/\lambda^2] = 1/\lambda$$

$$F_X(x/X \geq 2) = P(X \leq x/X \geq 2) = P(X \leq x, X \geq 2)/P(X \geq 2)$$

$$= P(X \leq x, X \geq 2)/P(X \geq 2)$$

$$= \frac{P(2 \leq X \leq x)}{P(X \geq 2)}$$

$$= \frac{F_X(x) - F_X(2)}{P(X \geq 2)}$$

$$\Rightarrow f_X(x/X \geq 2) = \frac{f_X(x)}{P(X \geq 2)}\ for\ x \geq 2$$
$$= 0,\ \text{otherwise}$$

$$P(X \geq 2) = \int_2^{\infty} f_X(x)dx = \int_2^{\infty}\lambda e^{-\lambda x}dx = \frac{\lambda e^{-\lambda x}}{-\lambda}\,\underset{2}{\infty} = -e^{-\lambda x} = e^{-2\lambda}$$

$$\Rightarrow f_X(x/X \geq 2) = \lambda e^{-\lambda x}e^{2\lambda}\ for\ x \geq 2$$
$$= 0,\ \text{otherwise}$$

$$E(X/X \geq 2) = \int_{-\infty}^{\infty} x f_X(x/X \geq 2) dx =$$

$$\int_2^{\infty} x\lambda e^{-\lambda x} e^{2\lambda} dx = \lambda e^{2\lambda} \int_2^{\infty} x \; d\left[\frac{e^{-\lambda x}}{-\lambda}\right]$$

$$= \lambda e^{2\lambda}\left[x\left[\frac{e^{-\lambda x}}{-\lambda}\right]_2^{\infty} - \int_2^{\infty} \frac{e^{-\lambda x}}{-\lambda} dx\right] = \lambda e^{2\lambda}\left[(2e^{-2\lambda}/\lambda) + e^{-2\lambda}/\lambda^2)\right]$$

$$= 2 + (1/\lambda)$$

Example 2.16. Let X $= [X1\ X2\ X3]^T$ is a three-dimensional zero-mean Gaussian random vector with covariance matrix C given by

$$\begin{matrix} & 1 & 3 & 0 \\ C = & 3 & 2 & 3 \\ & 0 & 3 & 1 \end{matrix}$$

The joint density function $f_{X1X2X3}(x1, x2, x3)$ is given as follows

$$f_{X1X2X3}(x1, x2, x3) = f_{\underline{X}}(\underline{x})$$
$$= \frac{1}{(2\Pi)^{\frac{n}{2}} |C|^{\frac{1}{2}}} e^{-\left(\frac{1}{2}\right)\left(\left[\underline{X} - \underline{\mu_X}\right]^T C^{-1} \left[\underline{X} - \underline{\mu_X}\right]\right)}$$

where $\underline{X} = [x1\ x2\ x3]^T$, $|c| = 16$, $n = 3$, $\underline{\mu_X} = [0\ 0\ 0]^T$ and

$$C^{-1} = \begin{bmatrix} 7 & 3 & -9 \\ 3 & -1 & 3 \\ -9 & 3 & 7 \end{bmatrix} * \left(\frac{1}{16}\right)$$

$$\Rightarrow f_{X1X2X3}(x1, x2, x3)$$
$$= \frac{1}{(2\Pi)^{\frac{3}{2}} |16|^{\frac{1}{2}}} e^{-\left(\frac{1}{2}\right)\left(\underline{[x1\ x2\ x3]} \begin{bmatrix} 7 & 3 & -9 \\ 3 & -1 & 3 \\ -9 & 3 & 7 \end{bmatrix} * \left(\frac{1}{16}\right) \begin{bmatrix} x1 \\ x2 \\ x3 \end{bmatrix}\right)}$$

$$\Rightarrow f_{X1X2X3}(x1, x2, x3)$$
$$= \frac{1}{(2\Pi)^{\frac{3}{2}} |16|^{\frac{1}{2}}} e^{-\left(\frac{1}{32}\right)(7\,x1^2 - x2^2 + 7\,x3^2 + 6\,x1\,x2 + 6\,x2\,x3 - 18 x1\,x3]}$$

If Y $=$ X1$+$X2$+$X3, then $f_Y(y)$ is Gaussian with mean m_Y and covariance matrix C_Y as shown below

$$Y = [1\ 1\ 1]\begin{bmatrix} x1 \\ x2 \\ x3 \end{bmatrix} \quad Y = A\underline{X},\ \text{where } A = [1\ 1\ 1]$$

$$\text{The covariance matrix } C_Y = A\, C_X A^T = [1\ 1\ 1]\begin{bmatrix} 1 & 3 & 0 \\ 3 & 2 & 3 \\ 0 & 3 & 1 \end{bmatrix}\begin{bmatrix} 1 \\ 1 \\ 1 \end{bmatrix} = 16$$

$$\text{Mean vector } m_Y = A\underline{m_X} = A\underline{\mu_X} = [1\ 1\ 1][0\ 0\ 0]^T = 0$$

Thus $f_Y(y) = \frac{1}{\sqrt{2\Pi * 13}} e^{-\frac{[y]^2}{2*16}}$

Example 2.17. Let $[X1\ X2]^T$ be the two-dimensional zero-mean Gaussian random vector with covariance matrix C given by $C = \begin{bmatrix} 4 & 2 \\ 2 & 4 \end{bmatrix}$. The conditional density functions $f_{X1/X2}(x1)$ and $f_{X2/X1}(x2)$ are also Gaussian distributed with the following specifications.

(i) $f_{X2/X1}(x2)$ is Gaussian with mean $= m2 - c21\ \mathrm{c11}^{-1}(x1 - m1)$ and variance $= c22 - c21\ \mathrm{c11}^{-1}c12$, where $[m1\ m2]^T$ is the mean vector of the two-dimensional Gaussian random vector in general. Also $C = \begin{bmatrix} \mathrm{c11} & c12 \\ c21 & c22 \end{bmatrix}$ be the corresponding generalized co-variance matrix.
In our case $\mathrm{m1} = 0$ and $\mathrm{m2} = 0$. $\mathrm{c11} = 4$ $\mathrm{c12} = 2$ $\mathrm{c21} = 2$ $\mathrm{c22} = 4$
There fore $f_{X2/X1}(x2)$ is Gaussian distributed with *mean* $= \left(-\frac{2}{4}\right) x1$ *and variance* $= 4 - \left(\frac{2}{4}\right) * 2 = 3$

$$f_{X2/X1}(x2) = \frac{1}{\sqrt{\Pi * 6}} e^{-\frac{\left[x2+\frac{1}{2}x1\right]^2}{6}}$$

(ii) $f_{X1/X2}(x1)$ is Gaussian with mean $= m1 - c12\ \mathrm{c22}^{-1}(x2 - m1)$ and variance $= \mathrm{c11} - c12\ \mathrm{c22}^{-1}c21$, where $[m1\ m2]^T$ is the mean vector of the two-dimensional Gaussian random vector in general. Also $C = \begin{bmatrix} \mathrm{c11} & c12 \\ c21 & c22 \end{bmatrix}$ be the corresponding generalized co-variance matrix.
In our case $\mathrm{m1} = 0$ and $\mathrm{m2} = 0$. $\mathrm{c11} = 4$ $\mathrm{c12} = 2$ $\mathrm{c21} = 2$ $\mathrm{c22} = 4$
There fore $f_{X2/X1}(x2)$ is Gaussian distributed with *mean* $= \left(-\frac{2}{4}\right) x1$ *and variance* $= 4 - \left(\frac{2}{4}\right) * 2 = 3$

$$f_{X2/X1}(x2) = \frac{1}{\sqrt{\Pi * 6}} e^{-\frac{\left[x2+\frac{1}{2}x1\right]^2}{6}}$$

Example 2.18. Let X1 and X2 are jointly Gaussian with zero-mean. Let the co-variance matrix of the random vector [X1 X2] is given as C $= \begin{bmatrix} 1 & \rho \\ \rho & 1 \end{bmatrix}$. Let Y1 = X1 + X2, Y2 = X1 − X2, then the co-variance matrix of the random vector [Y1 Y2] is given as

$$C_Y = AC_X A^T \begin{bmatrix} 1 & 1 \\ 1 & -1 \end{bmatrix} \begin{bmatrix} 1 & \rho \\ \rho & 1 \end{bmatrix} \begin{bmatrix} 1 & 1 \\ 1 & -1 \end{bmatrix} = \begin{bmatrix} 2(1+\rho) & 0 \\ 0 & 2(1-\rho) \end{bmatrix}$$

This implies the random variable Y1 and Y2 are uncorrelated. As the distribution is Gaussian, this also implies that the two random variables Y1 and Y2 are independent.

Similarly, for the case of co-variance matrix C $= \begin{bmatrix} \sigma 1^2 & \rho\sigma 1\sigma 2 \\ \rho\sigma 1\sigma 21 & \sigma 2^2 \end{bmatrix}$ and Y1 $= \frac{X1}{\sigma 1} + \frac{X2}{\sigma 2}$, Y2 $= \frac{X1}{\sigma 1} - \frac{X2}{\sigma 2}$, the co-variance matrix $C_Y = AC_X A^T$ is computed as following

$$\begin{bmatrix} 1/\sigma 1 & 1/\sigma 2 \\ 1/\sigma 1 & -1/\sigma 2 \end{bmatrix} \begin{bmatrix} \sigma 1^2 & \rho\sigma 1\sigma 2 \\ \rho\sigma 1\sigma 2 & \sigma 2^2 \end{bmatrix} \begin{bmatrix} 1/\sigma 1 & 1/\sigma 1 \\ 1/\sigma 2 & -1/\sigma 2 \end{bmatrix}$$

$$= \begin{bmatrix} \sigma 1 + \rho\sigma 1 & \sigma 2 + \rho\sigma 2 \\ \sigma 1 - \rho\sigma 1 & \rho\sigma^2 - \sigma 2 \end{bmatrix} \begin{bmatrix} 1/\sigma 1 & 1/\sigma 1 \\ 1/\sigma 2 & -1/\sigma 2 \end{bmatrix} = \begin{bmatrix} 2(1+\rho) & 0 \\ 0 & 2(1-\rho) \end{bmatrix}$$

This implies the random variable Y1 and Y2 are uncorrelated. As the distribution is Gaussian, this also implies that the two random variables Y1 and Y2 are independent.

2.29 Complex Random Variables

Consider the probability space (S,F,P), then the mapping of the outcome s ∈ sample space S to the complex line is called complex random variable (Fig. 2.24).

The complex random variable 'X' is represented as $X_r + j\ X_i$

$$\mathrm{E(X)} = \mathrm{E}(X_r) + j\ E(X_i)$$

The joint pdf of 'X_r' and 'X_i' is represented as $f_{X_r X_i}(x_r, x_i)$, which is the joint density function of the random variable 'X_r' and 'X_i'.

$$\mathrm{E(g(X))} = \mathrm{E(g}(X_r)) + j\ E(g(X_i))$$

The complex random variable X $= X_r + j\ X_i$ is defined as proper complex random variable if it satisfies the following condition.

Fig. 2.24 Illustration of the Complex random variables

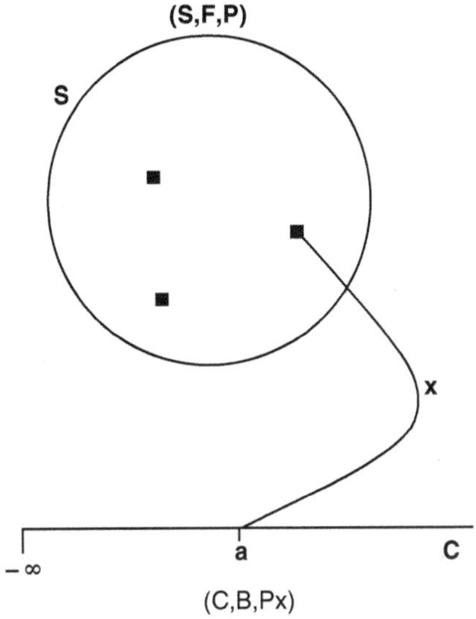

$$E(X^2) = 0$$

$$E(X^2) = E\left((X_r + j\ X_i)(X_r + j\ X_i)\right) = E(Xr^2 - Xi^2) + 2jE(XiXr)$$

$$\Rightarrow E(Xr^2) - E(Xi^2) = 0$$

$$(i.e)\ E(Xr^2) = E(Xi^2)$$

$$Also,\ E(Xi\ Xr) = 0$$

(i.e.) The random variable X_r *and* X_i are uncorrelated

The second moment of the random variables 'X_r' and 'X_i' are zero.

It can also be shown that for the proper complex random variable, variance of the random variables 'X_r' and 'X_i' are equal and they are uncorrelated.

E $((X - E(X))^2)$ is called pseudo covariance. The co-variance for the complex random variable 'X' is defined as E((X-E(X)) $(X - E(X))^H$)

2.30 Sequence of the Number and Its Convergence

Let the sequence of random variables be represented as $x_1 x_2 x_3 \ldots$, then the sequence converges to the constant x is represented as follows

$$\lim_{n\to\infty} x_n = x$$

For example $\left(1+\frac{1}{1}\right), \left(1+\frac{1}{2}\right), \left(1+\frac{1}{3}\right), \ldots \left(1+\frac{1}{n}\right)$ is the sequence of random variables.

$$\lim_{n\to\infty}\left(1+\frac{1}{n}\right)=1$$

Definition for convergence:

$$\lim_{n\to\infty} x_n = x$$

⇒ For any $\in > 0$, there exists N, such that $|x_n - x| < \in$ for all $n \geq N$

2.31 Sequence of Functions and Its Convergence

Let the sequence of functions be represented as $f_1(t)f_2(t)f_3(t)\ldots f_n(t)$, then the function converges to the function f(t) is represented as follows

- Point wise convergence $\lim_{n\to\infty} f_n(t) = f(t)$ for all 't'
- Weeker convergence $\lim_{n\to\infty} f_n(t) = f(t)$ for all t except at certain discrete time instants $t_1 t_2 t_3 \ldots t_n$
- Mean square convergence
  The sequence of functions converges to the function f(t) if it satisfies the following condition

$$\lim_{n\to\infty}\int_{-\infty}^{\infty} |f_n(t) - \mathrm{f}(t)|^2 \mathrm{dt} = 0$$

This can be interpreted as the sequence of numbers and they converge to the constant 0.

Example: The Fourier series representation of the periodic signal converges in mean square sense.

(i.e.) The function $f_n(t) = \sum_{k=-n}^{k=n} c_k e^{-j*2*pi*k*f0*t}$ converges to the function f(t) in mean square sense.

2.32 Sequence of Random Variable

Random variable is the function of mapping of the outcomes of the experiment (i.e.) events's ∈ the sample space S to the real line. Hence it comes under sequence of functions.

The sequence of functions $X_1(s)X_2(s)X_3(s)\ \ldots\ X_n(s)$ converges to the function X(s) (Another random variable) as follows

- Point wise convergence
  $\lim_{n\to\infty} X_n(s) = X(s)$ for all s
- Almost sure convergence
  $\lim_{n\to\infty} X_n(s) = X(s)$ for all s $\in \Omega$, where Ω is the field (i.e.) $\mathrm{p}(\Omega) = 1$
- Mean square convergence

$$\lim_{n\to\infty} X_n(s) = X(s)$$

If $\lim_{n\to\infty} E((X_n(s) - X(s))^2) = 0$
Note that $E((X_1(s) - X(s))^2)$, $E((X_2(s) - X(s))^2)$, $E((X_3(s) - X(s))^2), \ldots$ $E((X_n(s)-X(s))^2)$ can be viewed as the sequence of numbers and they converge to the constant 0.

- Convergence in probability
$\lim_{n\to\infty} X_n(s) = X(s)$ in probability sense if the sequence of numbers $P(|X_n - X| > \in)$ converges to the value 0 (i.e.) $\lim_{n\to\infty} P(|X_n - X| > \in) = 0$ for any $\in > 0$
- Convergence in distribution
$\lim_{n\to\infty} X_n(s) = X(s)$ in distribution sense if distribution function $F_{Xn}(\alpha) = F_X(\alpha)$ for all values of 'α' where $F_X(\alpha)$ is continuous.

Properties of the different types of convergence

1. If the sequence of random variable converges in mean square sense then they will converge in probability sense (Fig. 2.25)

$$P((|X_n - X| \geq \in) \geq \frac{E((X_n - X)^2)}{\in^2}$$

Proof. If the sequence of random variable converges in the mean square sense, $E((X_n - X)^2) = 0 \Rightarrow P((|X_n - X| \geq \in) \leq 0$. Probability cannot be less than zero and hence $P((|X_n - X| \geq \in) = 0$ and hence proved

2. If the sequence of random variable converges in almost sure sense then they will converge in probability sense. (Obvious from the definition)
3. If the sequence of random variable converges in probability sense then they will converge in distribution sense and hence

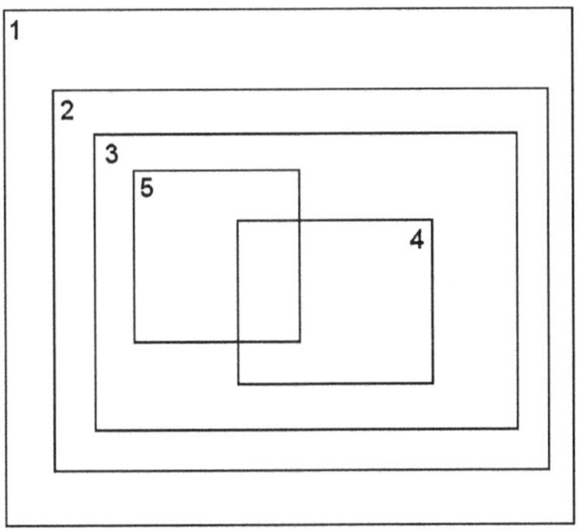

Fig. 2.25 Convergence of the Sequence of random variables
1. Strictly convergence
2. Convergence in distribution
3. Convergence in probability
4. Convergence in mean square sense
5. Convergence in almost sure sense

- If the sequence of random variable converges in mean square sense then they will converge in distribution sense.
- If the sequence of random variable converges in almost sure sense then they will converge in distribution sense.

2.33 Example for the Sequence of Random Variable

Suppose $X_1\ X_2 \ldots$ is the sequence of random variable such that $\mathrm{E}(X_i) = k$ *for all* i, $\mathrm{E}((X_i - k)^2)$ *is finite for all* i and the covariance $\mathrm{E}((X_i - k)(X_j - k)) = 0$ *for* $i \neq j$, then the sequence of random variable defined as $Y_1 Y_2 Y_3\ \ldots\ Y_n$ converges to the constant 'k' in mean square sense.

Where $Y_n = \left(\frac{1}{n}\right) \sum_{k=1}^{n} X_i$.

2.34 Central Limit Theorem

If we have independent and identically distributed random variables $X_1\ X_2 \ldots$ with mean 'm' and variance less than infinity, then the sequence of random variable $\left(\frac{1}{\sqrt{n}}\right) \sum_{i=1}^{n} (X_i - m)$ converges to the random variable X in distribution sense such that the random variable 'X' is having Gaussian probability density function with mean zero and constant variance.

Chapter 3
Random Process

3.1 Introduction

The mapping of the experimental outcomes s ∈ sample space S to the set of Random vectors is called as random process (Fig. 3.1). Individual random vector can be treated as the signal which varies as the function of time. (i.e.) Thus the random process can also be viewed as the mapping of the outcomes s ∈ sample space S to the set of signals as the function of time.

Example 3.1. An experiment has four equally likely outcomes 0,1,2,3 (i.e.) S = $\{0, 1, 2, 3\}$. The random process X_t is defined as $X_t = cos(2 * PI * s * t)$ for all s ∈ S.

In the above example, the outcome of the experiment '0' is mapped to the random vector

[1 1 1 1 1 1 1 1 1 ... 1]

Similarly the outcome of the experiment '1', '2', '3' are mapped to the set of random vectors as shown below. [Note that resolution of the variable 't' is 1/1,000]

'1'->

[1.0000 1.0000 0.9999 0.9998 0.9997 0.9995 0.9993 0.9990 0.9987 0.9984...

'2'->

[1.0000 0.9999 0.9997 0.9993 0.9987 0.9980 0.9972 0.9961 0.9950 0.9936 ...]

3-> [1.0000 0.9998 0.9993 0.9984 0.9972 0.9956 0.9936 0.9913 0.9887 0.9856 ...]

The mapped vectors are plotted as the function of time which are shown in the Fig. 3.1. Thus the random process can be viewed as the mapping of the outcomes of the experiment to the set of signals as the function of time.

E.S. Gopi, *Mathematical Summary for Digital Signal Processing Applications with Matlab*, DOI 10.1007/978-90-481-3747-3_3,

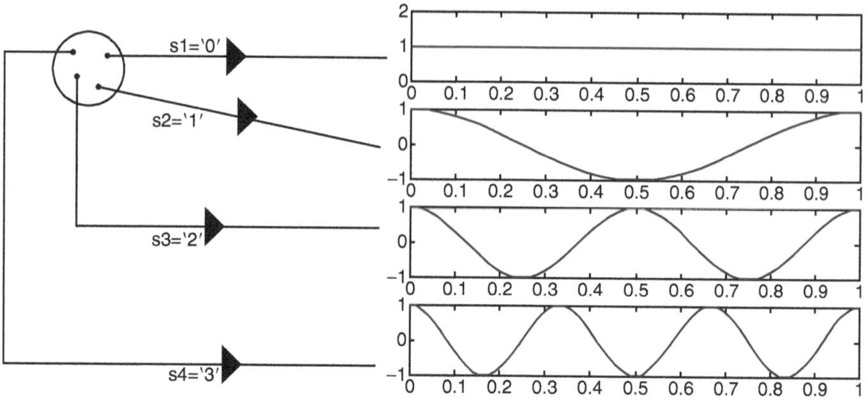

Fig. 3.1 Random process

3.2 Random Variable X_{t1}

The Random variable X_{t1} is obtained by sampling across the random process X_t at particular time instant 't1'. In the previous example the random variable X_0 $X_{0.25}$ $X_{0.5}$ are obtained by sampling the random process across the time instant '0', '0.25', '0.5' respectively as shown in the Fig. 3.2.

The random variable X_0 holds the values 0 with probability = 1
The random variable $X_{00.25}$ holds the values 1 with probability = 1/4
0 with probability = 1/2
−1 with probability = 1/4
The random variable $X_{0.5}$ holds the values 1 with probability = 1/2
−1 with probability = 1/2

3.3 Strictly Stationary Random Process with Order 1

Cumulative distribution function of the random variable $X_{t1+\tau}$ = Cumulative distribution function of the random variable X_{t1} for all values of τ. (i.e.) $F_{Xt1}(\alpha) = F_{Xt1+\tau}(\alpha)$ for all τ.

3.4 Strictly Stationary Random Process with Order 2

$$F_{Xt1,Xt2}(\alpha,\beta) = F_{Xt1+\tau,\ Xt2+\tau}(\alpha,\beta) \text{ for all } \tau, \text{t1}, \text{t2}, (\alpha,\beta)$$

If $\beta \xrightarrow{\textit{tends to}} \infty$, order 1 stationarity holds

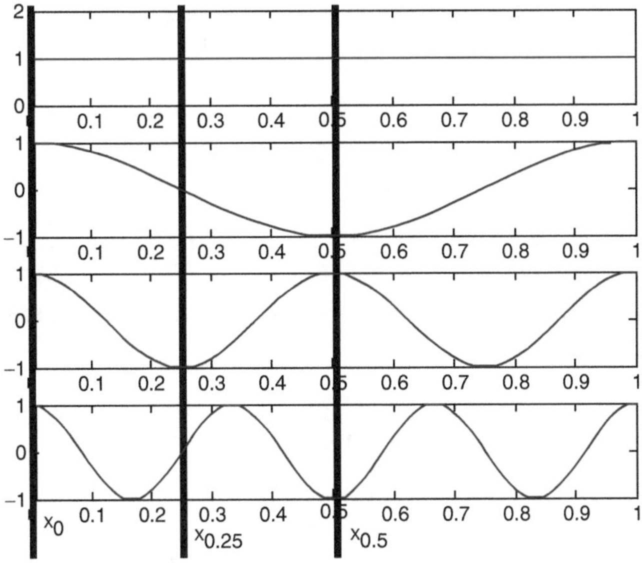

Fig. 3.2 Random variable X_{t1}

In general if the random process is strictly stationary at order n, it is strictly stationary for order n − 1.

Example 3.2. (a) X_n: Functions of outcomes of tossing the coin. If the Probability of Head is same for every toss, then the random process X_n is strictly stationary of order 1.

(b) Example 2.1 is the non-stationary random process.

(c) $X_t = A\ \cos(2^*\Pi^*f^*t + \Theta)$, '$\Theta$' is the random variable independent of 't' and is uniformly distributed between 0 to $2^*\Pi$. This is strictly stationary process.

3.5 Wide Sense Stationary Random Process

The Random process X_t is said to be wide-sense stationary random process if it satisfies the following conditions

- $m_X(t) = E(X_t) = \mathit{constant}$ for all time instant 't'.
- $E[X_t X_s) = R_X(t, s)$ is the function of t − s.

(i.e.) $E[X_t X_s) = R_X(t, s) = R_X(t - s) = R_X(\tau)$ for all values of 't' and 's'.

Example 3.3. $X_t = \mathrm{A}\ \cos(2^*\Pi^*\mathrm{f}^*\mathrm{t} + \Theta)$, '$\Theta$' is uniformly distributed between 0 to $2^*\Pi$.

$$\begin{aligned} m_X(t) &= \mathrm{E}(X_t) = E(\mathrm{A}\ \cos(2^*\Pi^*\mathrm{f}^*\mathrm{t} + \Theta) \\ &= \mathrm{A\ E}(\cos(2^*\Pi^*\mathrm{f}^*\mathrm{t} + \Theta) \end{aligned}$$

(Note Expectation is computed at one particular time instant 't' (i.e.) 't' is fixed. so the function $\cos(2^*\Pi^*\mathrm{f}^*\mathrm{t} + \Theta)$ is the function of the random variable 'Θ' only. (Represented as $\mathrm{g}(\Theta)$).

Let the probability density function of the random variable 'Θ' be represented as $\mathrm{f}(\Theta)$.

Therefore

$$\begin{aligned} &\mathrm{A\ E}(\cos(2^*\Pi^*\mathrm{f}^*\mathrm{t} + \Theta)) \\ &= \mathrm{A} \int g(\Theta) f(\Theta)\, \mathrm{d}\Theta \\ &= \mathrm{A} \int \cos(2*\Pi*\mathrm{f}*\mathrm{t} + \Theta)\ \mathrm{f}(\Theta)\ \mathrm{d}\Theta \\ &= \mathrm{A} \int_0^{2\Pi} \cos(2*\Pi*\mathrm{f}*\mathrm{t} + \Theta) \frac{1}{2*\Pi}\ \mathrm{d}\Theta \\ &= \frac{\mathrm{A}}{2*\Pi} \int_0^{2\Pi} \cos(2*\Pi*\mathrm{f}*\mathrm{t} + \Theta)\ \mathrm{d}\Theta \\ &= 0 \end{aligned}$$

$\therefore m_X(t)$ is Constant

Also Autocorrelation is computed as follows

$$\begin{aligned} &R_X(s,t) = R_X(t+\tau,t) = \mathrm{E}(X_{t+\tau} X_t) \\ &= \mathrm{E}(\mathrm{A\ E}(\cos(2^*\Pi^*\mathrm{f}^*(\mathrm{t}+\tau) + \Theta)\ \mathrm{A\ E}(\cos(2^*\Pi^*\mathrm{f}^*\mathrm{t} + \Theta)) \\ &= \mathrm{E}[A^2 \cos(2^*\Pi^*\mathrm{f}^*(\mathrm{t}+\tau) + \Theta)\ \cos(2^*\Pi^*\mathrm{f}^*\mathrm{t} + \Theta)] \\ &= A^2\ \mathrm{E}\,[\cos(2^*\Pi^*\mathrm{f}^*(\mathrm{t}+\tau) + \Theta)\ \cos(2^*\Pi^*\mathrm{f}^*\mathrm{t} + \Theta)] \\ &= \frac{A^2}{2} \mathrm{E}[\cos(2^*\Pi^*\mathrm{f}^*(2\mathrm{t}+\tau) + 2\Theta) + \cos(2^*\Pi^*\mathrm{f}^*\tau)] \\ &= \frac{A^2}{2} \mathrm{E}[\cos(2^*\Pi^*\mathrm{f}^*(2\mathrm{t}+\tau) + 2\Theta)] + \frac{A^2}{2} E[\cos(2^*\Pi^*\mathrm{f}^*\tau)] \end{aligned}$$

Note that $\cos(2^*\Pi^*\mathrm{f}^*\tau)$ is constant as τ and f are constant

$$\text{So } \frac{A^2}{2} E[\cos(2^*\Pi^*\mathrm{f}^*\tau)] = \frac{A^2}{2} \cos(2^*\Pi^*\mathrm{f}^*\tau)$$

I term

$\frac{A^2}{2}$E[cos(2*Π*f*(2t + τ) + 2Θ)] = 0 [Refer the steps involved in calculating $m_X(t)$]

$\therefore R_X(s,t) = R_X(t+\tau,s) = \frac{A^2}{2}\cos(2^*\Pi^*f^*\tau) = R_X(s-t) = R_X(\tau)$ is the function of the difference u − s = τ.

Thus X_t = A cos(2*Π*f*t + Θ), where 'Θ' is the random variable which is uniformly distributed between 0 to 2*Π is the wide sense stationary process.

3.6 Complex Random Process

The autocorrelation of the complex random process is given as

$$R_X(t,s) = \mathrm{E}(X_t X_S^*).$$

If the Complex random process is Wide Sense Stationary process, then $R_X(t,s) = \mathrm{E}\left(X_t X_S^*\right) = R_X(t-s) = R_X(\tau)$ which is the function of difference t − s = τ

3.7 Properties of Real and Complex Random Process

1. $R_X(0) = E\left(X_t^2\right)$ for real random process
   $R_X(0) = E(|X_t|^2)$ for complex random process
2. $R_X(\tau) = R_X(-\tau)$ for real random process
   $R_X(\tau) = R_X^*(-\tau)$ for complex random process
   Also Real $[R_X(\tau)]$ = Real $[R_X(-\tau)]$ (Even symmetry)
   Imaginary $[R_X(\tau)]$ = −Imaginary $[R_X(-\tau)]$ (Odd symmetry)
3. $|R_X(\tau)| \le R_X(0)$ for both real and complex random process
   (i.e.) $\left|\mathrm{E}\left(X_{t+\tau}X_\tau^*\right)\right| \le E(|X_t|^2)$ for complex random process
   $|\mathrm{E}(X_{t+\tau}X_t)| \le E\left(X_t^2\right)$ for real random process

3.8 Joint Strictly Stationary of Two Random Process

Consider two random process 'X_t' and 'Y_t'. They are said to be jointly strictly stationary if it satisfies the following condition

$$F_{Xt1,Yt2}(\alpha,\beta) = F_{Xt1+\tau,\ Yt2+\tau}(\alpha,\beta) \text{ for all } \tau, \mathrm{t1}, \mathrm{t2}, (\alpha,\beta).$$

Note that the two random process 'X_t' and 'Y_t' are individually stationary.

3.9 Jointly Wide Sense Stationary of Two Random Process

Consider two random process 'X_t' and 'Y_t'. They are said to be jointly Wide Sense Stationary (W.S.S) if it satisfies the following condition

1. $m_X(t) = E(X_t) = constant$ for all time instant 't'
2. E$[X_t X_s) = R_X(t, s)$ is the function of t – s.
   (i.e.) E$[X_t X_s) = R_X(t, s) = R_X(t - s) = R_X(\tau)$ for all values of 't' and 's'
3. $m_Y(t) = E(Y_t) = constant$ for all time instant 't'
4. E$[Y_t Y_s) = R_Y(t, s)$ is the function of t – s.
   (i.e.) E$[Y_t Y_s) = R_Y(t, s) = R_Y(t - s) = R_Y(\tau)$ for all values of 't' and 's'
5. E$[X_t Y_s] = R_{XY}(t, s) = R_{XY}(t - s) = R_{XY}(\tau)$ for all values of 't' and 's'

Note

(a) E$[X_t Y_s] = R_{XY}(t, s)$ is called as cross correlation function.
(b) $R_{XY}(t, s) = R_{YX}(s, t)$ for real random process
    $R_{XY}(\tau) = R_{YX}(-\tau)$ for W.S.S. real random process
(c) $R_{XY}(t, s) = R^*_{YX}(s, t)$ for Complex random process $R_{XY}(\tau) = R^*_{YX}(-\tau)$ for W.S.S. Complex random process

3.10 Correlation Matrix of the Random Column Vector $\begin{matrix} X_t \\ Y_s \end{matrix}$ for the Specific 't' 's'

$$\mathbf{E}\left(\begin{bmatrix} X_t \\ Y_s \end{bmatrix} |X_t \quad Y_s|\right) = \begin{bmatrix} E\left(X_t^2\right) & \mathrm{E}[X_t Y_s] \\ \mathrm{E}[Y_t X_s] & E\left(Y_t^2\right) \end{bmatrix}$$

3.11 Ergodic Process

Let 's1', 's2' ...'sn' be the outcomes of the experiment. Let $X_t(\mathrm{s1})$ be the signal as the function of time which is the map of the experiment S1. Similarly $X_t(\mathrm{s2})$ be the signal corresponding to the experiment S2. The set of functions as the outcomes of all the experiments forms the random process which is represented as X_t (see Fig. 3.3). Also let X_{t1} be the random variable which holds the values obtained by collecting the values across the process at some arbitrary time instant 't1'.

Ensemble average of the random variable X_{t1} is computed across the process and is given by

$$\int_{-\infty}^{\infty} x f_{X_{t1}}(x) dx$$

where f_{Xt1} is the probability density function of the random variable 'X_{t1}'.

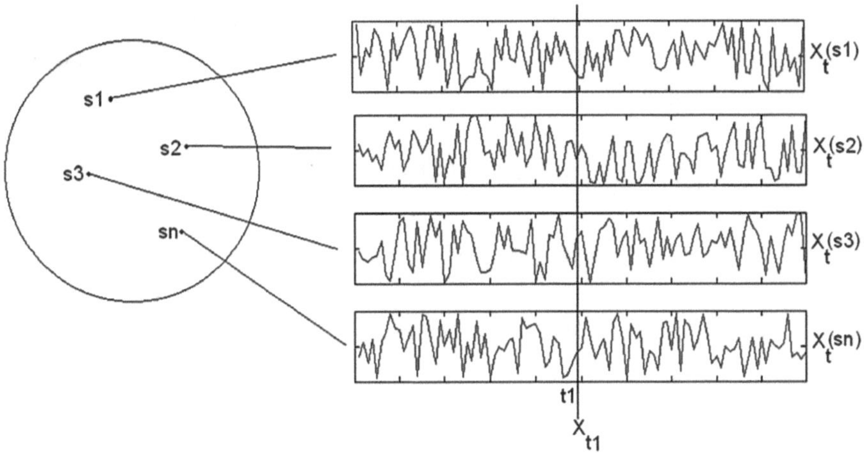

Fig. 3.3 Illustrations of Ergodic process

Time average of the arbitrary mapped signal $X_t(\mathrm{s1})$ is computed as follows

$$\lim_{T\to\infty}\left(\frac{1}{2T}\right)\int_{-T}^{T} X_t(\mathrm{s1})\,dt$$

In case of Wide Sense Stationary process, the ensemble average $\mathrm{E}(X_{t1})$ is constant.. If the above mentioned Ensemble average (constant) is equal to the Time average computed for any arbitrary mapped signal $X_t(\mathrm{s1})$ (say),the random process is called Ergodic in mean.

(i.e.) The Random process X_t is said to be Ergodic in mean if

$$\lim_{T\to\infty}\left(\frac{1}{2T}\right)\int_{-T}^{T} X_t(\mathrm{si})\,dt = \int_{-\infty}^{\infty} x f_{X_{t1}}(x)dx = \text{constant for all 'i'}.$$

Note that Random process X_t must be the W.S.S. process if it is Ergodic process. Similarly the random process is said be Ergodic in Auto correlation if the random process satisfies for the following condition. Let X_t be the W.S.S process. Ensemble average in auto correlation computation is given as

$$R_X(\tau) = \mathrm{E}(X_{t+\tau}X_t) = \iint xy\ f_{X_{t+\tau}X_t}(x,y)dx\,dy$$

Time average in auto correlation computation is given as

$$\lim_{T\to\infty}\left(\frac{1}{2T}\right)\int_{-T}^{T} X_t(\mathrm{s1})X_{t+\tau}(\mathrm{s1})\,dt$$

If Ensemble average is equal to the time average in auto correction computation, the random process is said to be Ergodic in autocorrelation. (i.e.) The random process X_t is said to be Ergodic in autocorrelation if

$$\lim_{T\to\infty}\left(\frac{1}{2T}\right)\int_{-T}^{T} X_t(\mathrm{si})X_{t+\tau}(\mathrm{si})\,dt$$

$$= \iint xy\ f_{X_{t+\tau}X_t}(x,y)dx\,dy = R_X(\tau) = \text{function of '}\tau\text{' for all 'i'}$$

Example 3.4. $X_t = \mathrm{A}\cos(2^*\Pi^*\mathrm{f}^*\mathrm{t} + \Theta)$, '$\Theta$' is uniformly distributed between 0 to $2^*\Pi$.

One particular map corresponding to the experimental outcome S1 is given as

$$X_t(\mathrm{s1}) = \mathrm{A}\ \cos(2^*\Pi^*\mathrm{f}^*\mathrm{t} + \Theta(\mathrm{s1}))$$

Ensemble average is given as $\mathrm{E}(X_t) = 0$ (see Example 3.3)

Time average

$$\lim_{T\to\infty}\left(\frac{1}{2T}\right)\int_{-T}^{T} X_t(\mathrm{s1})\,dt$$

$$= \lim_{T\to\infty}\left(\frac{1}{2T}\right)\int_{-T}^{T} \mathrm{A}\ \cos(2*\Pi*\mathrm{f}*\mathrm{t} + \Theta(\mathrm{s1}))\,dt$$

$$= \lim_{T\to\infty}\left(\left(\frac{1}{2T}\right)\mathrm{A}\sin(2*\Pi*\mathrm{f}*\mathrm{T} + \Theta(\mathrm{s1}))/(2*\Pi*\mathrm{f})\right)$$

A $\sin(2*\Pi*\mathrm{f}*\mathrm{T} + \Theta(\mathrm{s1}))/(2*\Pi*\mathrm{f}))$ is bounded between two constants. Say between 'M1' and 'M2'.

$$\text{(i.e.)}\quad \lim_{T\to\infty}\left(\left(\frac{1}{2T}\right)M1 \le \lim_{T\to\infty}\left(\left(\frac{1}{2T}\right)\mathrm{A}\ \sin(2*\Pi*\mathrm{f}*\mathrm{T} + \Theta(\mathrm{s1}))/\ (2*\Pi*\mathrm{f})\right)\right.$$

$$\le \lim_{T\to\infty}\left(\left(\frac{1}{2T}\right)M2 \Rightarrow 0\right.$$

$$\le \lim_{T\to\infty}\left(\left(\frac{1}{2T}\right)\mathrm{A}\ \sin(2*\Pi*\mathrm{f}*\mathrm{T} + \Theta(\mathrm{s1}))/(2*\Pi*\mathrm{f})\right) \le 0$$

$$\Rightarrow \lim_{T\to\infty}\left(\left(\frac{1}{2T}\right) \text{A} \sin(2 * \Pi * \text{f} * \text{T} + \Theta(\text{s1}))/(2 * \Pi * \text{f})\right) = 0$$

$$\Rightarrow \lim_{T\to\infty}\left(\frac{1}{2T}\right)\int_{-T}^{T} X_t(\text{s1})\,dt = 0$$

Thus Ensemble average = Time average = constant = 0 and hence the random process $X_t = \text{A}\cos(2^*\Pi^*\text{f}^*\text{t} + \Theta)$ (where 'Θ' is uniformly distributed between 0 to $2^*\Pi$) is Ergodic in mean. In the same manner, it can be shown that $X_t = \text{A}\cos(2^*\Pi^*\text{f}^*\text{t} + \Theta)$ (where 'Θ' is uniformly distributed between 0 to $2^*\Pi$) is Ergodic in autocorrelation as shown below.

Ensemble average in auto correlation is given as $R_X(\tau) = \text{E}(X_{t+\tau}X_t) = \frac{A^2}{2}\cos(2^*\Pi^*\text{f}^*\tau)$ (see Example 2.3)

Time average in auto correlation is computed as

$$\begin{aligned}
&\lim_{T\to\infty}\left(\frac{1}{2T}\right)\int_{-T}^{T} X_t(\text{s1})X_{t+\tau}(\text{s1})\,dt \\
&= \lim_{T\to\infty}\left(\frac{1}{2T}\right)\int_{-T}^{T} \text{A}\cos(2 * \Pi * \text{f} * (\text{t}+\tau) + \Theta(\text{s1})) \\
&\quad \times \text{A}\cos(2 * \Pi * \text{f} * \text{t} + \Theta(\text{s1}))\,dt \\
&= \lim_{T\to\infty}\left(\frac{1}{2T}\right)\int_{-T}^{T} (\text{A}^2/2)[\cos(2 * \Pi * \text{f} * (2\text{t}+\tau) + 2\,\Theta(\text{s1})) \\
&\quad + \cos(2 * \Pi * \text{f} * \tau)]\,dt
\end{aligned}$$

First term

$$\lim_{T\to\infty}\left(\frac{1}{2T}\right)\int_{-T}^{T} (\text{A}^2/2)[\cos(2 * \Pi * \text{f} * (2\text{t}+\tau) + 2\,\Theta(\text{s1}))]\,dt = 0$$

(As described in Time average in mean)

$$\begin{aligned}
&\Rightarrow \lim_{T\to\infty}\left(\frac{1}{2T}\right)\int_{-T}^{T} X_t(\text{s1})X_{t+\tau}(\text{s1})\,dt \\
&= \lim_{T\to\infty}\left(\frac{1}{2T}\right)\int_{-T}^{T} (\text{A}^2/2)[\cos(2 * \Pi * \text{f} * (2\text{t}+\tau) + 2\,\Theta(\text{s1})) \\
&\quad + \cos(2 * \Pi * \text{f} * \tau)]\,dt \\
&= \lim_{T\to\infty}\left(\frac{1}{2T}\right)\int_{-T}^{T} (\text{A}^2/2)\,[\cos(2 * \Pi * \text{f} * \tau)]\,dt \\
&= \lim_{T\to\infty}\left(\frac{1}{2T}\right)(\text{A}^2/2)[\cos(2 * \Pi * \text{f} * \tau)]\int_{-T}^{T} dt
\end{aligned}$$

$$= \lim_{T\to\infty}\left(\frac{1}{2T}\right)(A^2/2)[\cos(2*\Pi*f*\tau)](2T)$$
$$= \lim_{T\to\infty}(A^2/2)[\cos(2*\Pi*f*\tau)]$$
$$= (A^2/2)[\cos(2*\Pi*f*\tau)]$$

Thus Ensemble average in autocorrelation = Time average in auto correlation

$$= (A^2/2)[\cos(2*\Pi*f*\tau)] \text{ is the function of } `\tau'.$$

Hence the random process $X_t = A\cos(2^*\Pi^*f^*t + \Theta)$ (where 'Θ' is uniformly distributed between (0 to 2*Π) is Ergodic in autocorrelation

3.12 Independent Random Process

Let X_t and Y_t be two random processes. Let $\underline{X} = [X_{t1}\ X_{t2}\ X_{t3}\ X_{t4} \ldots\ X_{tn}]$ be the random vector obtained by sampling across the random process X_t at time instants t1, t2, ... tn. Similarly the random vector $\underline{Y} = [Y_{t1}\ Y_{t2}\ Y_{t3}\ Y_{t4} \ldots\ Y_{tn}]$ is obtained by sampling across the random process Y_t.

The random process X_t and Y_t are independent if the random vectors $\underline{X}$ and $\underline{Y}$ are independent random vectors.

$$\text{(i.e.) } F_{\underline{XY}}() = F_{\underline{X}}()F_{\underline{Y}}()$$

3.13 Uncorrelated Random Process

Let X_t and Y_t be two random processes. Let $\underline{X} = [X_{t1}\ X_{t2}\ X_{t3}\ X_{t4} \ldots\ X_{tn}]$ be the random vector obtained by sampling across the random process X_t at time instants t1, t2,... tn. Similarly the random vector $\underline{Y} = [Y_{t1}\ Y_{t2}\ Y_{t3}\ Y_{t4} \ldots\ Y_{tn}]$ is obtained by sampling across the random process Y_t.

The random process X_t and Y_t are uncorrelated if the cross-covariance matrix computed as E $((\underline{X} - m_{\underline{X}})^T\ (\underline{Y} - m_{\underline{Y}}))$ is the diagonal matrix.

3.14 Random Process as the Input and Output of the System

Consider the Linear time invariant system described by its impulse response h(t) (Fig. 3.4). Let X_t be the W.S.S. random process which is given as the input to the system h(t) and Y_t be the corresponding output random process which is also W.S.S. Then $\int_{-\infty}^{\infty} h(\tau)X_{t-\tau}\,d\tau$ converges to the output random process in mean square sense. (i.e.)

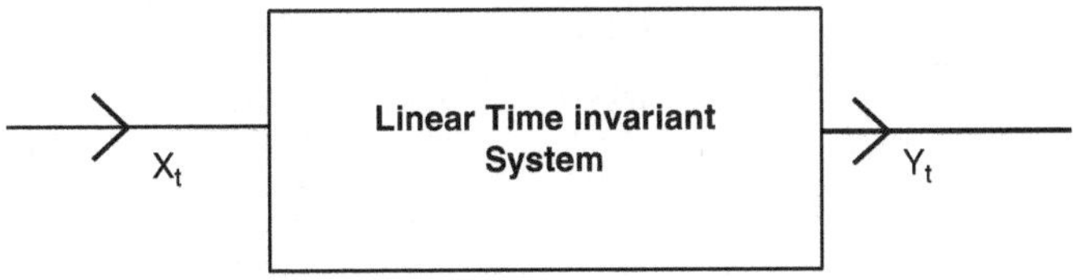

Fig. 3.4 Random process as the input and output of the LTI system

$$Y_t{}^{m.s.s} = \int_{-\infty}^{\infty} h(\tau)X_{t-\tau}d\tau$$
$$\Rightarrow \mathrm{E}([Y_t - \int_{-\infty}^{\infty} h(\tau)X_{t-\tau}d\tau]^2) = 0$$

Properties

1. Mean of the output random process $E(Y_t)$ is constant

   Proof.

$$\begin{aligned} E(Y_t) &= E\left(\int_{-\infty}^{\infty} h(\tau)X_{t-\tau}d\tau\right. \\ &= \int_{-\infty}^{\infty} h(\tau)E\left(X_{t-\tau}\right)d\tau \\ &= \int_{-\infty}^{\infty} h(\tau)m_X\, d\tau \\ &= m_X\int_{-\infty}^{\infty} h(\tau)\, d\tau \\ &= \text{constant.} \end{aligned}$$

2. $R_{YX}\,(t,\ s) = \mathrm{E}(Y_t X_s] = R_{YX}(\tau) = \mathrm{h}(\tau) * R_X(\tau)$

   Proof.

$$\begin{aligned} R_{YX}(t,s) &= \mathrm{E}(Y_t X_s] = R_{YX}(\tau) \\ &= E\left[\left(\int_{-\infty}^{\infty} h(\tau)X_{t-\tau}d\tau\right) X_s\right] \\ &= \int_{-\infty}^{\infty} h(\tau)E(X_{t-\tau}X_s)\, d\tau \\ &= \int_{-\infty}^{\infty} h(\tau)R_X(t-\tau-s)\, d\tau \\ &= \int_{-\infty}^{\infty} h(\tau)R_X(t-s-\tau)\, d\tau \\ &= \int_{-\infty}^{\infty} h(\tau)R_X(\gamma-\tau)\, d\tau \end{aligned}$$

$$\gamma = t - s(say)$$
$$\Rightarrow R_{YX}(\tau) = \mathrm{h}(\tau) * R_X(\tau)$$

3. $R_Y(\tau) = R_X(\tau)^*\mathrm{h}(\tau) * \mathrm{h}(-\tau)$

$$\begin{aligned}
R_Y(t,s) &= \mathrm{E}(Y_t Y_s] = R_Y(\tau) \\
&= E\left[Y_t\left(\int_{-\infty}^{\infty} h(\tau)X_{s-\tau}d\tau\right)\right] \\
&= \int_{-\infty}^{\infty} h(\tau)E(Y_t X_{s-\tau})\,d\tau \\
&= \int_{-\infty}^{\infty} h(\tau)R_{YX}(t-s+\tau)\,d\tau \\
&= \int_{-\infty}^{\infty} h(\tau)R_{YX}(\gamma+\tau)\,d\tau \\
\gamma &= t - s(say)
\end{aligned}$$

Let $\delta = -\tau$

$$\Rightarrow R_Y(\tau) = \int_{-\infty}^{\infty} h(-\delta)R_{YX}(-\delta+\tau)\,d\tau$$
$$= \int_{-\infty}^{\infty} h'(\delta)R_{YX}(-\delta+\tau)\,d\tau$$

Let

$$\begin{aligned}
h'(\delta) &= \mathrm{h}(-\delta)(\mathrm{say}) \\
&\Rightarrow R_Y(\tau) = h'(\tau) * R_{YX}(\tau) \\
&\Rightarrow R_Y(\tau) = h(-\tau) * R_{YX}(\tau)
\end{aligned}$$

We know,

$$\begin{aligned}
R_{YX}(\tau) &= \mathrm{h}(\tau) * R_X(\tau) \\
&\Rightarrow R_Y(\tau) = h(-\tau) * \mathrm{h}(\tau) * R_X(\tau)
\end{aligned}$$

Note that the above mentioned properties are true for the complex random process also.

3.15 Power Spectral Density (PSD)

The power spectral density of the W.S.S. random process X_t is defined as the Fourier transformation of its autocorrelation function $R_X(\tau)$. (i.e.)

$$S_X(f) = \int_{-\infty}^{\infty} R_X(\tau)\, e^{-j2\Pi f\tau} d\tau$$

Consider the Linear time invariant system described by its impulse response h(t). Let X_t be the W.S.S. random process which is given as the input to the system h(t) and Y_t be the corresponding output random process which is also W.S.S.

We have shown that

$$R_Y(\tau) = h(-\tau) * \mathrm{h}(\tau) * R_X(\tau)$$

Taking Fourier transformation on both sides

$$\Rightarrow S_Y(f) = H(f)H(-f)S_X(f)$$
$$\Rightarrow S_Y(f) = H(f)H^*(f)S_X(f) (\text{Assuming h}(\tau) \text{ is the real function})$$
$$\Rightarrow S_Y(f) = |H(f)|^2 S_X(f)$$

We have also shown that

$$R_{YX}(\tau) = \mathrm{h}(-\tau) * R_X(\tau)$$

Taking Fourier transformation on both sides

$$\Rightarrow S_{YX}(f) = H(f)S_X(f)$$

The power spectral density $S_{YX}(f)$ is called as Cross power spectral density.

Properties of power spectral density

1. $S_X(0) = \int_{-\infty}^{\infty} R_X(\tau)\, d\tau$
2. $R_X(0) = \int_{-\infty}^{\infty} S_X(f)\, d\tau = E\left(X_t{}^2\right)$
   Note that mean square value is obtained from all frequencies of the spectral density
3. $S_X(f)$ is always real for all values of 'f' and ≥ 0

$$\begin{aligned} S_X(f) &= \int_{-\infty}^{\infty} R_X(\tau)\, e^{-j2\Pi\tau} d\tau \\ &= \int_{-\infty}^{0} R_X(\tau)\, e^{-j2\Pi f\tau} d\tau + \int_{0}^{\infty} R_X(\tau)\, e^{-j2\Pi f\tau} d\tau \\ &= \int_{-\infty}^{0} R_X(\tau)\, e^{-j2\Pi\tau} d\tau + \int_{0}^{\infty} R_X(\tau)\, e^{-j2\Pi\tau} d\tau \end{aligned}$$

Let u $= -\tau$

$$= \int_{0}^{\infty} R_X(-u)\, e^{j2\Pi f u} du + \int_{0}^{\infty} R_X(\tau)\, e^{-j2\Pi f\tau} d\tau$$

$$= \int_0^{\infty} R_X(-u)\, e^{j2\Pi f u} du + \int_0^{\infty} R_X(\tau)\, e^{-j2\Pi f \tau} d\tau$$

$$= \int_0^{\infty} R_X^*(u)\, e^{j2\Pi f u} du + \int_0^{\infty} R_X(\tau)\, e^{-j2\Pi f \tau} d\tau$$

[Using the property of complex auto correlation (i.e) $R_X(-u) = R_X^*(u)$]

$$= \int_0^{\infty} R_X^*(u)\, e^{j2\Pi f u} du + \int_0^{\infty} R_X(\tau)\, e^{-j2\Pi f \tau} d\tau$$

$$\int_0^{\infty} [R_X(\tau) e^{-j2\Pi f \tau} d\tau]^* \int_0^{\infty} R_X(\tau)\, e^{-j2\Pi f \tau} d\tau$$

which is the real value and hence power spectral density $S_X(f)$ is always positive. (i.e.) $S_X(f) \geq 0$

4. If the W.S.S. random process X_t is real, then $S_X(f) = S_X(-f)$
5. Consider the W.S.S. random process X_t and the corresponding autocorrelation function and spectral density function are given as $R_X(\tau)$ and $S_X(f)$ respectively and $S_X\ (f) = 0$ for |f| > W, then the random process X_t is said to be band limited with Bandwidth 'W'
6. Consider the system which is Band limited with bandwidth 'W' (Fig. 3.5). Consider the band limited W.S.S. random process X_t which is given as the input to the system. The output of the system is the random process Y_t, then $\mathrm{E}([X_t - Y_t)^2) = 0$

Suppose W_t and V_t are the responses of the system $H_1(f)$ and $H_2(f)$ to the band limited process 'X_t' (Fig. 3.6)

If $H_1(f) = H_2(f)$ for all $|f| \leq W$, then W_t and V_t are equal in Mean Square Sense. (i.e.) $E[(W_t - Y_t)^2] = 0$

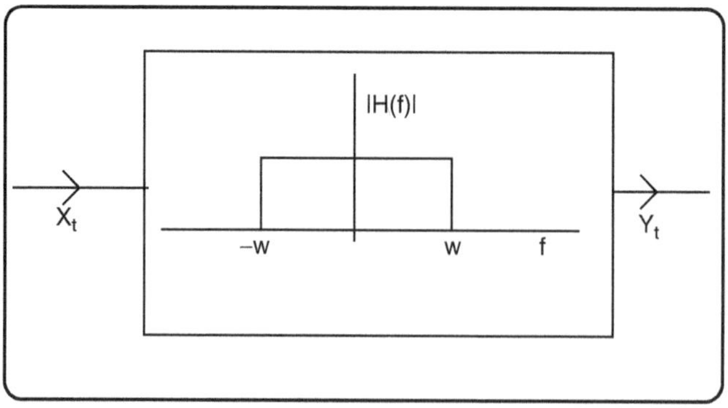

Fig. 3.5 Transfer function of the band limited system

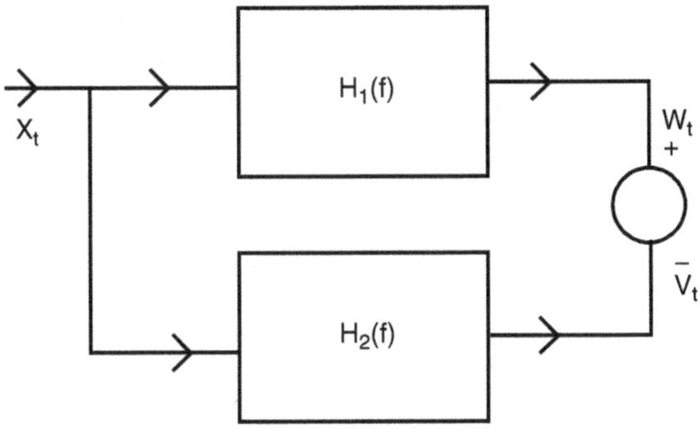

Fig. 3.6 Response of the system to the band limited process

Proof. Let $Z_t = (W_t - Y_t)$

$$\Rightarrow E\left[(W_t - Y_t)^2\right] = E\left[{Z_t}^2\right]$$

We know that,

$$S_Z(f) = S_X(f)|H(f)|^2$$
$$where\ H(f) = H_1(f) - H_2(f)$$

Also $E\left[{Z_t}^2\right] = R_z(0)$

$$= \int_{-\infty}^{\infty} S_Z(f)\, df$$
$$= \int_{-\infty}^{\infty} S_X(f)|H(f)|^2\, df$$

Because H (f) = 0 for all $|f| \le W$ and $S_X(f) = 0\, for\ all\, |f| > W$
Thus $E[(W_t - Y_t)^2] = 0$

3.16 White Random Process (Noise)

The random process X_t is said to be White Gaussian Random process, if Mean = $m_X\ (t) = E(X_t) = 0$ for all time instant 't'

$$R_X(t,s) = R_X(\tau)\delta(\tau)\frac{N_o}{2}$$
$$S_X(f) = constant.$$

White noise has zero mean and infinite variance, which cannot be realized in practical situation. The white noise obtained in real time can be viewed as the filtered white noises through the Band limited filter whose frequency response flat over the bandwidth of interest.

3.17 Gaussian Random Process

The Random vector X_t is said to be Gaussian Random Process if the random vector obtained by sampling across the process X_t at time instants t1, t2, t3, ... tn (Represented as $[X_{t1}\ X_{t2}\ \ldots\ X_{tn}]$) is a jointly Gaussian random vector.

Properties

1. If the input to the Linear, stable system is Gaussian random process, then output is also Gaussian random process. Note that the system can also be Time variant
2. If X_t is Gaussian and W.S.S. it is also Strictly Stationary
3. The random process X_t is said to be White Gaussian Random process, if
   - X_t W.S.S. Gaussian Random process
   - Mean $= m_X(t) = E(X_t) = 0$ for all time instant 't'
   - $R_X(t,\ s) = R_X(\tau)\delta(\tau)\frac{N_o}{2}$
   - $S_X(f) = constant$

Note that there can be White Non-Gaussian Random process.

Example 3.5. **Gaussian Random process**

1. Wiener process [Non-stationary random process]
   - Mean $= m_X(t) = E(X_t) = 0$ for all time instant 't'
   - $R_X(t,s) = \sigma^2 \min(t,s) + m^2 ts$ for all t, s ≥ 0
2. Gauss-Markov process[Stationary random process]
   - $m_X(t) = 0$
   - $R_X(t,s) = \sigma^2 e^{-\beta|t-s|}, \sigma^2, \beta > 0$ for all t, s ≥ 0
   - Let t1,t2,t3 be the three samples instance of the Gauss Markov random process with t3 > t2 > t1, then $f_{Xt3/Xt2\ Xt1} = f_{Xt3/Xt2}$

In general conditional density of the random variable obtained at particular time instant tn given the random variables obtained at set of time instants t1,t2,t3,..tn − 1 with t1 < t2 < ...tn − 1, is equal to the conditional density function of the random variable obtained at time instant tn given the random variable obtained at time instant tn − 1.

(i.e.) Let t1,t2,t3,..tn be the three samples instance of the Gauss Markov random process with tn > tn−1 > tn−2...t2 > t1, then $f_{Xtn/Xtn-1\ Xtn-2\ \ldots\ Xt1} = f_{Xtn/Xtn-1}$

3.18 Cyclo Stationary Random Process

The random process X_t is said to be strictly stationary random process

$$\begin{aligned} &F_{Xt1,Xt2,Xt3,Xt4\ldots,Xtn}(\alpha 1, \alpha 2, \alpha 3, \ldots \alpha n) \\ &\quad = F_{Xt1+\tau,Xt2+\tau,Xt3+\tau,Xt4+\tau\ldots+Xtn+\tau}(\alpha 1, \alpha 2, \alpha 3, \ldots \alpha n) \end{aligned}$$

for all $\tau, \alpha 1, \alpha 2, \alpha 3, \ldots \alpha n$

If $\tau = \text{n T}$, where n = $\ldots -2,\ -1,\ 0,\ 1,\ 2\ \ldots$ and T is constant, the random process is said to be cyclo stationary random process with period 'T'.

Example 3.6. $X_t = \sum_{n=-\infty}^{\infty} A_n p(t - nT)$ is the cyclo stationary random process, *where* A_n is a discrete time strictly stationary process. p(t) is the function of 't'.

3.19 Wide Sense Cyclo Stationary Random Process

The random process X_t is W.S. Cyclo stationary random process if

$$\begin{aligned} m_X(t) &= 0 \\ R_X(t,s) &= R_X(t + nT, s + nT) \end{aligned}$$

Example 3.7. Consider the Discrete wide sense stationary random process A_n that takes the values +1 and −1 with equal probability at all time instants (Stream of Binary data). Let the pulse used to modulate the above mentioned binary stream is p(t) having nonzero values for $0 \le t \le T$. The random process defined as $X_t = \sum_{n=-\infty}^{\infty} A_n p(t - nT)$ is cyclo stationary.

Let this random process be the input to the channel input and the random process Y_t is the random process of the output of the channel (i.e.) in the receiver section. The random process Y_t is represented as follows. $Y_t = \sum_{n=-\infty}^{\infty} A_n p(t - nT - \Theta)$, where Θ be the time delay of the pulse p (t) which can be viewed as the random variable which is uniformly distributed between 0 to T. The random process Y_t is the Wide Sense Stationary random process.

Proof.

$$\begin{aligned} X_t &= \sum_{n=-\infty}^{\infty} A_n p(t - nT) \\ m_X(t) &= \mathrm{E}(X_t) = \sum_{n=-\infty}^{\infty} E(A_n) p(t - nT) \\ &= \sum_{n=-\infty}^{\infty} k p(t - nT) \text{ which is periodic with time} \\ &\qquad \text{period 'T'} \\ &\Rightarrow m_X(t + T) = m_X(t) \\ X_t &= \sum_{n=-\infty}^{\infty} A_n p(t - nT) \end{aligned}$$

$$R_X(t,s) = \mathrm{E}(X_t X_s) = \sum_{n=-\infty}^{\infty} \sum_{k=-\infty}^{\infty} E(A_n A_k) p(t-kT) p(s-nT)$$
$$= \sum_{n=-\infty}^{\infty} \sum_{k=-\infty}^{\infty} R_A(n-k) p(t-kT)\ p(s-nT)$$

Changing the limit m = n − k

$$= \sum_{n=-\infty}^{\infty} \sum_{k=-\infty}^{\infty} R_A\,(n-k)\,p\,(t-kT)\ \ p\,(s-nT)$$
$$= \sum_{n=-\infty}^{\infty} \sum_{m=-\infty}^{\infty} R_A(m)\,p(t-nT+mT)\ \ p\,(s-nT)$$
$$\Rightarrow R_X\,(t+T, s+T) = \sum_{n=-\infty}^{\infty} \sum_{m=-\infty}^{\infty} R_A\,(m)\,p\,(t+T-nT+mT)$$
$$p\,(s+T-nT)$$
$$= \sum_{n=-\infty}^{\infty} \sum_{m=-\infty}^{\infty} R_A\,(m)\,p\,(t-nT+mT)\ \ p\,(s-nT) = R_X\,(t,s)$$

Hence $X_t = \sum_{n=-\infty}^{\infty} A_n p(t-nT)$ is cyclo stationary process.

The random process $Y_t = \sum_{n=-\infty}^{\infty} A_n p(t-nT-\Theta)$ [Θ is uniformly distributed between 0 to T] is wide sense stationary random process

Proof. **Computation of mean**

$$Y_t = \sum_{n=-\infty}^{\infty} A_n p(t-nT-\Theta)\ \mathrm{i}$$
$$\mathrm{E}(Y_t) = E_\Theta[E(Y_t/\Theta = a)]$$
$$= E_\Theta\,[E(Y_t/\Theta = a)]$$
$$[E(Y_t/\Theta = a)]$$
$$= \sum_{n=-\infty}^{\infty} E(A_n) p(t-nT-a)$$
$$= \sum_{n=-\infty}^{\infty} kp(t-nT-a)$$
$$= m_X(t-a)[\textit{Because}\ m_X(t) = \sum_{n=-\infty}^{\infty} kp(t-nT)$$

which is periodic with time period 'T'

$$\Rightarrow E_\Theta[E(Y_t/\Theta = a)]$$
$$= E_\Theta[m_X(t-a)]$$
$$= \left(\frac{1}{T}\right)\int_0^T m_X(t-a)\,da$$

Note that $m_X(t)$ is periodic with time period 'T' $.m_X(t-a)$ is the shifted version of $m_X(t)$ and hence $\left(\frac{1}{T}\right)\int_0^T m_X(t-a)da$ is constant.

$$E(Y_{t+\tau}Y_t) = E_\Theta[E(Y_{t+\tau}Y_t/\Theta = a)]$$

$$\begin{aligned}E(Y_{t+\tau}Y_t/\Theta = a) &= \sum\nolimits_{n=-\infty}^{\infty}\sum\nolimits_{m=-\infty}^{\infty} A_n p(t+\tau-mT-a)\\ &\quad A_m p(t-nT-a)\\ &= \sum\nolimits_{n=-\infty}^{\infty}\sum\nolimits_{m=-\infty}^{\infty} E(A_nA_m)p(t+\tau-mT-a)\\ &\quad p(t-nT-a)\\ &= \sum\nolimits_{n=-\infty}^{\infty}\sum\nolimits_{m=-\infty}^{\infty} R_A(n-m)p(t+\tau-mT-a)\\ &\quad p(t-nT-a)\end{aligned}$$

Let k = n − m

$$\begin{aligned}&= \sum_{n=-\infty}^{\infty}\sum_{k=-\infty}^{\infty} R_A(k)p(t+\tau-nT+kT-a)\\ &\quad p(t-nT-a)\\ &= \sum_{n=-\infty}^{\infty}\sum_{k=-\infty}^{\infty} R_A(k)p(t+\tau-nT+kT-a)\\ &\quad p(t-nT-a)\\ &\therefore E_\Theta[E(Y_{t+\tau}Y_t/\Theta = a)]\\ &= \left(\frac{1}{T}\right)\int_0^T \sum\nolimits_{n=-\infty}^{\infty}\sum\nolimits_{k=-\infty}^{\infty} R_A(k)p(t+\tau-nT\\ &\quad +kT-a)p(t-nT-a)da\\ &= \left(\frac{1}{T}\right)\sum_{K=-\infty}^{\infty} R_A(k)\sum_{n=-\infty}^{\infty}\int_0^T p(t+\tau-nT\\ &\quad +kT-a)p(t-nT-a)da\end{aligned}$$

Let u = $t - nT$ − a

$$= \left(\frac{1}{T}\right)\sum_{K=-\infty}^{\infty} R_A(k)\sum_{n=-\infty}^{\infty}\int_{t-nT-T}^{t-nT} p(u+\tau+kT)p(u)du$$

Consider $\sum_{n=-\infty}^{\infty}\int_{t-nT-T}^{t-nT} p(u+\tau+kT)p(u)du$

$$\begin{aligned}\text{Let f(u)} &= p(u+\tau+kT)p(u)\\ &= \cdots + \int_{t+2T}^{t+3T} f(u)du + \int_{t+3T}^{t+4T} f(u)du + \ldots \int_{t-T}^{t} f(u)du\\ &\quad + \int_{t-2T}^{t-T} f(u)du + \int_{t-T}^{t-2T} f(u)du + \cdots\end{aligned}$$

$$= \int_{-\infty}^{\infty} f(u)du$$
$$= \int_{-\infty}^{\infty} p(u + \tau + kT)p(u)\,du$$
$$= \int_{-\infty}^{\infty} p(u + \tau + kT)p(u)\,du$$

$= R_P(\tau + kT)$ which is the autocorrelation of the deterministic signal pulse p(t).

$$\Rightarrow \left(\frac{1}{T}\right) \sum_{K=-\infty}^{\infty} R_A(k) \sum_{n=-\infty}^{\infty} \int_0^T p(t + \tau - nT + kT - a)p(t - nT - a)da$$

$$\Rightarrow R_Y(\tau) = \left(\frac{1}{T}\right) \sum\nolimits_{K=-\infty}^{\infty} R_A(k)R_P(\tau + kT)$$

Which is the function of $t - s = \tau$ and hence $Y_t = \sum_{n=-\infty}^{\infty} A_n p(t - nT - \Theta)$ ['Θ' is uniformly distributed between 0 to T.] is Wide Sense Stationary process.

Taking Fourier transform on both sides of the equation

$$R_Y(\tau) = \left(\frac{1}{T}\right) \sum\nolimits_{K=-\infty}^{\infty} R_A(k)R_P(\tau + kT)$$

we get

$$S_Y(f) = \left(\frac{1}{T}\right) \sum_{K=-\infty}^{\infty} R_A(k)e^{-j2*\Pi*f*k*T}|P(f)|^2$$

Where P(f) is the Fourier transformation of the autocorrelation function p(t).

Also note that $R_P(t) = p(t)^*p(-t)$ and hence Fourier transform of $R_P(\tau)$ is $P(f)|^2$

3.20 Sampling and Reconstruction of Random Process

Consider the Band limited continuous random process X_t with bandwidth 'W' (i.e.) $S_X(f) = 0\forall|f| > W$, which is sampled with sampling rate $\left(\frac{1}{T}\right) \geq 2W$ to obtain the discrete random process X_{nT} Then the sequence of random process ${X_t}^{(N)}$ converges to random process X_t in mean square sense as N–> ∞, where

$${X_t}^{(N)} = \sum\nolimits_{n=-N}^{n=N} X_{nT}\ sinc\left(\left(\frac{t}{T}\right) - n\right) .\text{(i.e.) E}\left(\left|X_t - {X_t}^{(N)}\right|^2\right)$$
$$= 0 \text{ as N-> } \infty.$$

Proof. The requirement is to show that $E\left(|X_t - X_t^{(N)}|^2\right) = 0$ as N−> ∞.

$$E\left(\left|X_t - X_t^{(N)}\right|^2\right) = E\left(\left|X_t - \sum_{n=-N}^{n=N} X_{nT}\ sinc\left(\left(\frac{t}{T}\right)-n\right)\right|^2\right)$$
$$= E\left(|X_t|^2\right) - \sum_{n=-N}^{N} E\left(X_t X_{nT}^*\right)\ sinc\left(\left(\frac{t}{T}\right)-n\right)$$
$$-\sum_{n=-N}^{N} E\left(X_{nt} X_t^*\right)\ sinc\left(\left(\frac{t}{T}\right)-n\right)$$
$$+\sum_{n=-N}^{N}\sum_{m=-N}^{N} E\left(X_{nT} X_{mT}^*\right) sinc\left(\left(\frac{t}{T}\right)-n\right)$$
$$sinc\left(\left(\frac{t}{T}\right)-m\right)$$

Note

Note that if X_t is bandlimited then the corresponding autocorrelation function (i.e.) $R_X(t)$ bandlimited and $R_X(t)$ can be viewed as the band limited signal and hence using sampling theorem, $R_X(t)$ can be reconstructed using the following formula

$$R_X(t) = \sum_{n=-\infty}^{\infty} R_X(nT))sinc\left(\left(\frac{t}{T}\right)-n\right)$$

The shifted version of the signal (i.e.) $R_X(t-\tau)$ can be reconstructed using the formula as shown below.

$$R_X(t-\tau) = \sum_{n=-\infty}^{\infty} R_X(nT-\tau))sinc\left(\left(\frac{t}{T}\right)-n\right)$$

At $t = \tau$

$$R_X(0) = \sum_{n=-\infty}^{\infty} R_X(nT-t))sinc\left(\left(\frac{t}{T}\right)-n\right) \text{---------}(1)$$

Similarly we can show that

$$R_X(0) = \sum_{n=-\infty}^{\infty} R_X(t-nT))sinc\left(\left(\frac{t}{T}\right)-n\right) \text{---------}(2)$$

Consider the four terms in the expanded form of the equation $E\left(|X_t - X_t^{(N)}|^2\right)$ (as shown above). Applying the limit N → ∞ individually on the four terms we get the following.

First term $E(|X_t|^2) = R_X(0)$

$$\text{Second term}: \sum_{n=-N}^{N} E(X_t X_{nT}^*)sinc\left(\left(\frac{t}{T}\right)-n\right)$$

$$\lim_{N\to\infty}\sum_{n=-N}^{N} R_X(t-nT))sinc\left(\left(\frac{t}{T}\right)-n\right) = R_X(0)$$

(From Eq. (2))

$$\text{Third term}: \sum\nolimits_{n=-N}^{N} E(X_{nT}X_t^*)sinc\left(\left(\frac{t}{T}\right)-n\right)$$

$$\lim_{N\to\infty}\sum_{n=-N}^{N} R_X(nT-t))sinc\left(\left(\frac{t}{T}\right)-n\right) = R_X(0)$$

(From Eq. (1))
Fourth term

$$\sum\nolimits_{n=-N}^{N}\sum\nolimits_{m=-N}^{N} E(X_{nT}X_{mT}^*)sinc\left(\left(\frac{t}{T}\right)-n\right)sinc\left(\left(\frac{t}{T}\right)-m\right)$$
$$=\sum\nolimits_{n=-N}^{N}\sum\nolimits_{m=-N}^{N} R_X(nT-mT)sinc\left(\left(\frac{t}{T}\right)-n\right)sinc\left(\left(\frac{t}{T}\right)-m\right)$$

Consider the term

$$\lim_{N\to\infty}\sum_{m=-N}^{N} R_X(nT-mT)sinc\left(\left(\frac{t}{T}\right)-n\right) = R_X(t-mT)$$

(From the Reconstruction formula of sampling theorem)

$$\lim_{N\to\infty}\sum_{m=-N}^{N} R_X(t-mT)sinc\left(\left(\frac{t}{T}\right)-m\right) = R_X(0)$$

(From Eq. (2))
Thus the

$$\lim_{N\to\infty} E\left(\left|X_t - X_t^{(N)}\right|^2\right) = E\left(\left|X_t - \sum\nolimits_{n=-N}^{n=N} X_{nT}sinc\left(\left(\frac{t}{T}\right)-n\right)\right|^2\right)$$
$$R_X(0) - R_X(0) - R_X(0) + R_X(0) = 0$$

Hence proved

3.21 Band Pass Random Process

The Random process is said to be Band pass Random process if its spectral density have band pass frequency response.

Example 3.8. The Random process $X_t \triangleq X_t{}^I \cos(2\Pi f_c t) - X_t{}^Q \sin(2\Pi f_c t)$ forms the Band pass W.S.S. Random process if the following conditions are satisfied.

- $X_t{}^I$ and $X_t{}^Q$ are the Low pass random process
- They are Jointly W.S.S. Process
- $R_{X^I}(\tau) = R_X Q(\tau)$
- $R_{X^I X^Q}(\tau) = -R_{X^Q X^I}(\tau)$

Proof. To make the random process X_t as the Wide sense stationary process, $R_X(t,u)$ should be the function of $\tau = \mathrm{t} - \mathrm{u}$

$$R_X(t+\tau,\tau) = E(X_{t+\tau}X_\tau)$$

Let A = $X_{t+\tau}{}^I \cos(2\Pi f_c(t+\tau)) - X_{t+\tau}{}^Q \sin(2\Pi f_c(t+\tau))$
B = $X_t{}^I \cos(2\Pi f_c t) - X_t{}^Q \sin(2\Pi f_c t)$
=E(AB) consists of four terms.
I term: E$(X_{t+\tau}{}^I \cos(2\Pi f_c(t+\tau)) X_t{}^I \cos(2\Pi f_c t))$

$$= \mathrm{E}\left(X_{t+\tau}{}^I X_t{}^I \frac{1}{2}\left[\cos(2\Pi f_c(2t+\tau)) + \cos(2\Pi f_c \tau)\right]\right)$$

$$= R_{X^I}(\tau)\frac{1}{2}\left[\cos\left(2\Pi f_c(2t+\tau)\right) + \cos(2\Pi f_c \tau)\right]$$

Second term: $-\mathrm{E}\left(X_{t+\tau}{}^I \cos(2\Pi f_c(t+\tau)) X_t{}^Q \sin(2\Pi f_c t)\right)$

$$= -\mathrm{E}\left(X_{t+\tau}{}^I X_t{}^Q \frac{1}{2}\left[\sin(2\Pi f_c(2t+\tau)) - \sin(2\Pi f_c \tau)\right]\right)$$

$$= -R_{X^I X^Q}(\tau)\frac{1}{2}\left[\sin(2\Pi f_c(2t+\tau)) - \sin(2\Pi f_c \tau)\right]$$

Third term: $-\mathrm{E}\left(X_{t+\tau}{}^Q \sin(2\Pi f_c(t+\tau)) X_t{}^I \cos(2\Pi f_c t)\right)$

$$= \mathrm{E}\left(X_{t+\tau}{}^Q X_t{}^I \frac{1}{2}\left[\sin(2\Pi f_c(2t+\tau)) + \sin(2\Pi f_c \tau)\right]\right)$$

$$= R_{X^I X^Q}(\tau)\frac{1}{2}\left[\sin(2\Pi f_c(2t+\tau)) + \sin(2\Pi f_c \tau)\right]$$

Fourth term: $\mathrm{E}\left(X_{t+\tau}{}^Q \sin(2\Pi f_c(t+\tau)) X_t{}^Q \sin(2\Pi f_c t)\right)$

$$= \mathrm{E}\left(X_{t+\tau}{}^Q X_t{}^Q \frac{1}{2}\left[-\cos(2\Pi f_c(2t+\tau)) + \cos(2\Pi f_c \tau)\right]\right)$$

$$= R_{X^Q}(\tau)\frac{1}{2}\left[-\cos(2\Pi f_c(2t+\tau)) + \cos(2\Pi f_c \tau)\right]$$

$$= \left[\frac{R_{X^I}(\tau) + R_{X^Q}(\tau)}{\mathbf{2}}\right]\cos(2\Pi f_c \tau) + \left[\frac{R_{X^I X^Q}(\tau) - R_{X^Q X^I}(\tau)}{\mathbf{2}}\right]\sin(2\Pi f_c \tau)$$

$$+\left[\frac{R_{X^I}(\tau)+R_{X^Q}(\tau)}{2}\right]\cos(2\Pi f_c(2t+\tau))$$
$$+\left[\frac{-R_{X^IX^Q}(\tau)-R_{X^QX^I}(\tau)}{2}\right]\sin(2\Pi f_c(2t+\tau))$$

To make the above expression independent of 't' First and second term is already independent of 't'. To make the third and fourth terms independent of 't', the following conditions have to be satisfied.

$$R_{X^I}(\tau)=R_{X^Q}(\tau)$$
$$-R_{X^IX^Q}(\tau)=R_{X^QX^I}(\tau)$$

Hence proved.

3.22 Random Process as the Input to the Hilbert Transformation as the System

Let W.S.S random process X_t be the input to the system (Hilbert transform) whose frequency response is as shown below (Figs. 3.7 and 3.8).

The Hilbert transform of the random process X_t is represented as $\widehat{X_t}$.

Properties

1. $S_{\hat{X}}(f)=S_X(f)$

$$S_{\hat{X}}(f)=S_X(f)|H(f)|^2$$

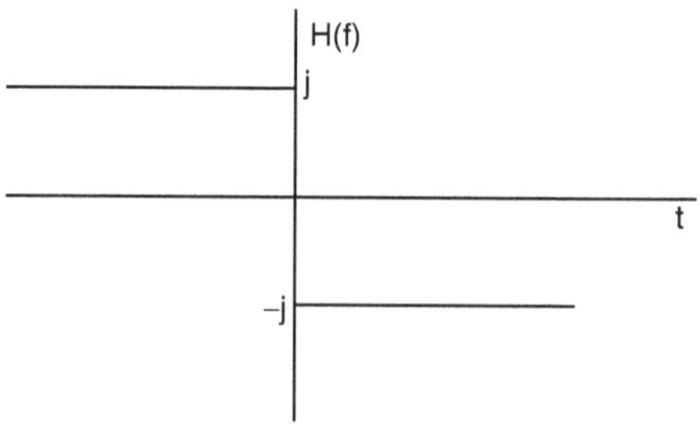

Fig. 3.7 Transfer function of the Hilbert transformation

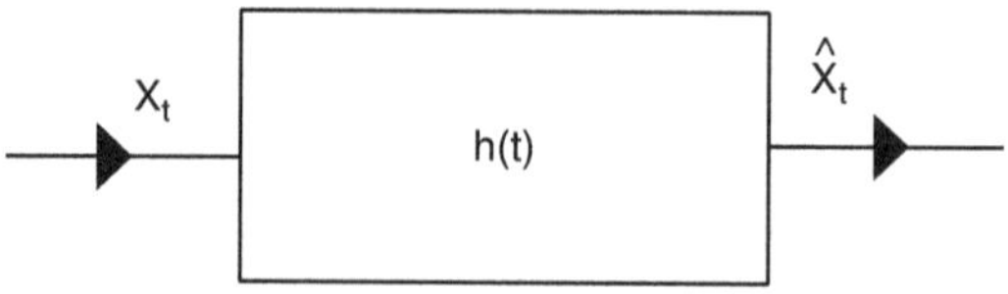

Fig. 3.8 Hilbert transformation system

$$|H(f)|^2 = 1 \text{ for all f (see Graph)}$$
$$\text{Hence } S_{\hat{X}}(f) = S_X(f)$$

2. $S_{X\hat{X}}(f) = -S_{\hat{X}X}(f)$
We know,

$$R_{\hat{X}X}(\tau) = R_X(\tau) * h(\tau)$$
$$\Rightarrow R_{\hat{X}X}(-\tau) = R_X(-\tau) * h(-\tau)$$

The function $R_X(\tau)$ is the even function. $\therefore$ $R_X(\tau) = R_X(-\tau)$
The impulse response of the Hilbert transformation system is given as

$$h(t) = \left(\frac{1}{\Pi t}\right) \textit{ for all t}$$
$$\Rightarrow R_{\hat{X}X}(-\tau) = -R_X(\tau) * h(\tau)$$
$$\Rightarrow -R_{\hat{X}X}(\tau)$$

Also we know $R_{\hat{X}X}(\tau) = \mathrm{E}(\widehat{X_{t+\tau}} X_t) = \mathrm{E}(X_t \widehat{X_{t+\tau}}) = R_{X\hat{X}}(-\tau)$

$$\Rightarrow R_{\hat{X}X}(-\tau) = R_{X\,\hat{X}}(\tau)$$

From the above

$$\Rightarrow R_{X\,\hat{X}}(\tau) = -R_{\hat{X}X}(\tau)$$

Taking Fourier transformation on both sides, we get

$$S_{X\hat{X}}(f) = -S_{\hat{X}X}(f)$$

3. $E(X_t \widehat{X_t}) = 0$

Proof. From property 2 we get,

$$R_{X\,\hat{X}}(\tau) = -R_{\hat{X}\,X}(\tau)$$
$$\Rightarrow R_{X\,\hat{X}}(0) = -R_{\hat{X}\,X}(0) - - - - - - (1)$$

Also we know

$$E(X_t \widehat{X_t}) = E(\widehat{X_t} X_t)$$
$$\Rightarrow R_{X\,\hat{X}}(0) = R_{\hat{X}\,X}(0) - - - - - - - (2)$$

From (1) and (2) we conclude $R_{X\hat{X}}(0) = 0$

$$\Rightarrow E(X_t \widehat{X_t}) = 0$$

4. Let $Z_t = X_t + j\widehat{X_t}$, then

$$\begin{aligned} \mathrm{E}(Z_{t+\tau}Z_t^{*}) &= \mathrm{E}((X_{t+\tau} + j\widehat{X_{t+\tau}})(X_{t+\tau} - j\widehat{X_{t+}})] \\ &= E(X_{t+\tau}X_t) - j\ E(X_{t+\tau}\widehat{X_t}) + jE(\widehat{X_{t+\tau}}X_t)E(\widehat{X_{t+\tau}\widehat{X_t}}) \\ &\Rightarrow R_Z(\tau) = R_X(\tau) - jR_{X\hat{X}}(\tau) + jR_{\hat{X}X}(\tau) + R_{\hat{X}}(\tau) \end{aligned}$$

From the property 1 and property 2, $R_X(\tau) = R_{\hat{X}}(\tau)$

$$\begin{aligned} R_{X\hat{X}}(\tau) &= -R_{\hat{X}X}(\tau) \\ &= R_X(\tau) + 2\mathrm{j}R_{\hat{X}X}(\tau) \\ &\Rightarrow R_Z(\tau) = 2R_X(\tau) + 2\mathrm{j}R_X(\tau) * h(\tau) \end{aligned}$$

In frequency domain (By taking Fourier Transform), we get,

$$\begin{aligned} S_Z(f) &= 2S_X(f) + 2jS_X(f)H(f) \\ &\Rightarrow S_Z(f) = 2S_X(f)(1 + \mathrm{jH(f)})\ \text{(See Figure 3-7)} \\ &\Rightarrow S_Z(f) = 4S_X(f)\ for\ f > 0 \\ &= 0\ for\ f < 0 \end{aligned}$$

3.23 Two Jointly W.S.S Low Pass Random Process Obtained Using W.S.S. Band Pass Random Process and Its Hilbert Transformation

Let X_t be the W.S.S. Band pass Random process

$$\text{Define } X_t^{+} \triangleq X_t + j\,\widehat{X_t}$$

From the property 4 of Hilbert transformation, we get

$$\begin{aligned} S_{X^+}(f) &= 4S_X(f)\, for\ f > 0 \\ &= 0\, for\ f < 0 \end{aligned}$$

Define $\tilde{X}_t \triangleq X_t{}^{+}e^{-j2*\Pi*f_c*t}$

$$\begin{aligned} E(\tilde{X}_{t+\tau}\,\tilde{X}_t) &= R_{X\sim}(\tau) = \mathrm{E}\left(X_{t+\tau}{}^{+}e^{-j2*\Pi*f_c*(t+\tau)}\ X_t{}^{+}e^{-j2*\Pi*f_c*(t)}\right) \\ &= E\left(X_{t-\tau}{}^{+}X_t{}^{+}e^{-j2*\Pi*f_c*(\tau)}\right) \\ &\Rightarrow R_{X\sim}(\tau) = R_{X+}(\tau)e^{-j2*\Pi*f_c*(\tau)} \\ &\Rightarrow S_{X\sim}(f) = S_{X+}(f + f_c) \end{aligned}$$

Let $X_t^{\sim} = X_t^{I} + j\ X_t^{Q}$

$$\Rightarrow X_t^{I} = Real\left(X_t^{+}e^{-j2*\Pi*f_c*(t)}\right) = Real\left(\left(X_t + j\widehat{X_t}\right)e^{-j2*\Pi*f_c*(t)}\right)$$
$$= X_t\cos\left(2*\Pi*f_c*t\right) + \widehat{X_t}\ sin\left(2*\Pi*f_c*t\right)$$

Similarly

$$\begin{aligned} X_t^{Q} &= Imaginary(X_t^{+}e^{-j2*\Pi*f_c*t}) \\ &= \widehat{X_t}\cos(2*\Pi*f_c*t) - X_t\ sin(2*\Pi*f_c*t) \end{aligned}$$

In the same fashion it can be shown that

$X_t = X_t^{I}\cos(2*\Pi*f_c*t) - X_t^{Q}\sin(2*\Pi*f_c*t)$ (Which is of the same form as mentioned in Example 2.11)

The random process X_t^{I} and X_t^{Q} satisfies the following conditions.

- $R_{X^I}(\tau) = R_{X^Q}(\tau)$
- $R_{X^I X^Q}(\tau) = -R_{X^Q X^I}(\tau)$
- $S_{X^I}(f) = S_{X^Q}(f) = \left(\frac{1}{4}\right)[S_{X^{\sim}}(f) + S_{X^{\sim}}(-f)]$
  (Low pass frequency response)
- $S_{X^I X^Q}(f) = \left(\frac{j}{4}\right)[S_{X^{\sim}}(f) - S_{X^{\sim}}(-f)]$
  (Low pass frequency response)

Thus jointly wide sense stationary low pass random process X_t^{I} and X_t^{Q} are generated using the Bandpass random process X_t and its Hilbert transformation $\widehat{X_t}$ as mentioned below

$$\begin{aligned} X_t^{I} &= X_t\cos(2*\Pi*f_c*t) + \widehat{X_t}\ \sin(2*\Pi*f_c*t) \\ X_t^{Q} &= \widehat{X_t}\cos(2*\Pi*f_c*t) - X_t\ \sin(2*\Pi*f_c*t) \end{aligned}$$

Proof. $R_{X^I}(\tau) = E\left(X_{t+\tau}^{I}\left(X_t^{I}\right)^*\right)$

We know

$$\begin{aligned} X_t^{\sim} &= X_t^{I} + j\ X_t^{Q} \\ &\Rightarrow X_t^{I} = \frac{(X_t^{\sim} + (X_t^{\sim})^*)}{2} \\ &\Rightarrow E(X_{t+\tau}^{I}(X_t^{I})^*) = E\left(\frac{(X_{t+\tau}^{\sim} + (X_{t+\tau}^{\sim})^*)}{2}\frac{(X_t^{\sim} + (X_t^{\sim})^*)}{2}\right) \\ &= \frac{1}{4}E(X_{t+\tau}^{\sim}X_t^{\sim}) + \frac{1}{4}E(X_{t+\tau}^{\sim}(X_t^{\sim})^*) + \frac{1}{4}E((X_{t+\tau}^{\sim})^*X_t^{\sim}) \\ &\quad + \frac{1}{4}E((X_{t+\tau}^{\sim})^*(X_t^{\sim})^*) \end{aligned}$$

I term: $\frac{1}{4}E\left(X_{t+\tau}^{\sim}X_t^{\sim}\right) = 0$

$$= \frac{1}{4}E\left(X_{t+\tau}{}^{+}e^{-j2*pi*f_c(t+\tau)}X_t{}^{+}e^{-j2*pi*f_c\tau}\right)$$

$$= \frac{1}{4}E\left(\left(X_{t+\tau} + j\ \widehat{X_{t+\tau}}\right)e^{-j2*pi*f_c(t+\tau)}\left(X_t + j\ \widehat{X_t}\right)e^{-j2*pi*f_c\tau}\right)$$

$$= e^{-j2*pi*f_c(2t+\tau)}\frac{1}{4}E\left((X_{t+\tau} + j\widehat{X_{t+\tau}})\ \left(X_t + j\ \widehat{X_t}\right)\right)$$

$$= e^{-j2*pi*f_c(2t+\tau)}\frac{1}{4}\left(E(X_{t+\tau}X_t) + jE\left(\widehat{X_{t+\tau}}X_t\right) + jE\left(X_{t+\tau}\ \widehat{X_t}\right)\right.$$
$$\left. - E(\widehat{X_{t+\tau}}\widehat{X_t}\right)$$

$$= e^{-j2*pi*f_c(2t+\tau)}\frac{1}{4}\left(R_X(\tau) + \mathrm{j}R_{X\hat{X}}(\tau) + \mathrm{j}R_{\hat{X}X}(\tau) - R_{\hat{X}}(\tau)\right)$$

From the property of Hilbert transformation mentioned in 3.22, we get the following

$$R_X(\tau) = R_{\hat{X}}(\tau)$$
$$R_{X\hat{X}}(\tau) = -R_{\hat{X}X}(\tau)$$

Hence $E(X_{t+\tau}{}^{\sim}X_t{}^{\sim}) = 0$

II term: $\frac{1}{4}E(X_{t+\tau}{}^{\sim}(X_t{}^{\sim})^*) = \frac{1}{4}R_{X^\sim}(\tau)$

III term: $\frac{1}{4}E((X_{t+\tau}{}^{\sim})^*X_t{}^{\sim}) = \frac{1}{4}E(X_t{}^{\sim}(X_{t+\tau}{}^{\sim})^*) = \frac{1}{4}R_{X^\sim}(-\tau)$

IV term: $\frac{1}{4}E((X_{t+\tau}{}^{\sim})^*(X_t{}^{\sim})^*) = 0$

This can be obtained in the same fashion as that of the I term.

Thus

$$R_{X^I}(\tau) = \frac{1}{4}\left[R_{X^\sim}(\tau) + R_{X^\sim}(-\tau)\right]$$

Taking Fourier transformation on both sides we get

$$S_{X^I}(\mathrm{f}) = \frac{1}{4}[S_{X^\sim}(\mathrm{f}) + S_{X^\sim}(-f)]$$

In the same fashion we can show

$$S_{X^Q}(\mathrm{f}) = \frac{1}{4}[S_{X^\sim}(\mathrm{f}) + S_{X^\sim}(-f)]$$

$$and\ S_{X^IX^Q}(f) = \left(\frac{j}{4}\right)[S_{X^\sim}(f) - S_{X^\sim}(-f)]$$

Example 3.9. Consider the Band pass random process X_t whose spectral density is as shown below (Figs. 3.9 and 3.10).

$$R_{X^\sim}(\tau) = E(X_{t+\tau}{}^{\sim}(X_t{}^{\sim})^*) = E\left(X_{t+\tau}{}^{+}(X_t{}^{+})^*e^{-j2*pi*f_c(t+\tau)}e^{j2*pi*f_c(t)}\right)$$
$$= R_{X^+}(\tau)e^{-j2*pi*f_c(\tau)}$$
$$\Rightarrow S_{X^\sim}(\mathrm{f}) = S_{X^+}(\mathrm{f} + \mathrm{f_c})$$

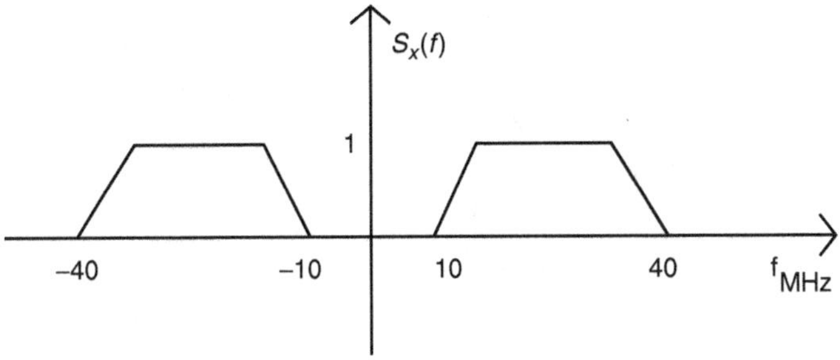

Fig. 3.9 Spectral density of the band pass random process

Fig. 3.10 Spectral density of the X_t^+

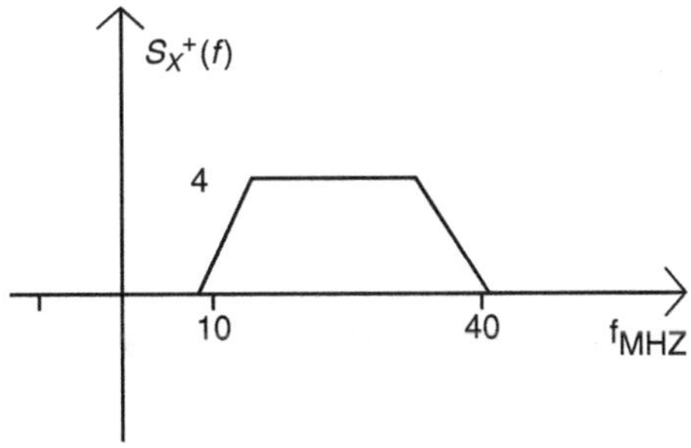

Fig. 3.11 Spectral density of the X_t^+

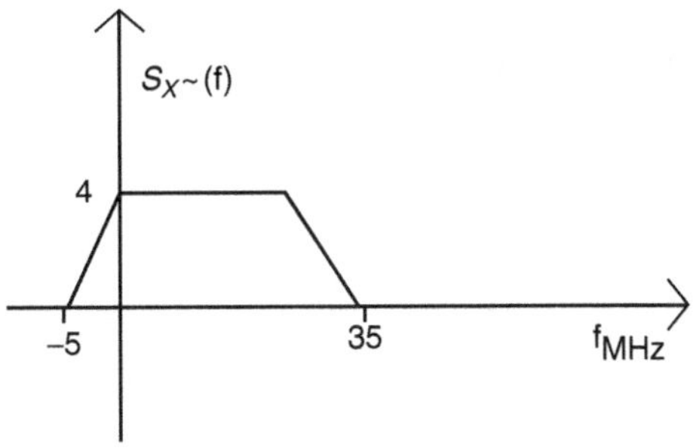

We know that

$$S_X + (\mathrm{f}) = 4S_X(f)\,for\ f > 0$$
$$= 0\,for\ f < 0$$

Let f_c be chosen as 15 MHz so that the spectral density $S_{X\sim}$(f) looks like the following (Fig. 3.11).

Thus the spectral density of the X_t^I and the X_t^Q are as shown below (Figs. 3.12 and 3.13).

Note that S_{X^I}(f), S_{X^Q}(f) and −j $S_{X^I X^Q}(f)$ are having low pass characteristics.

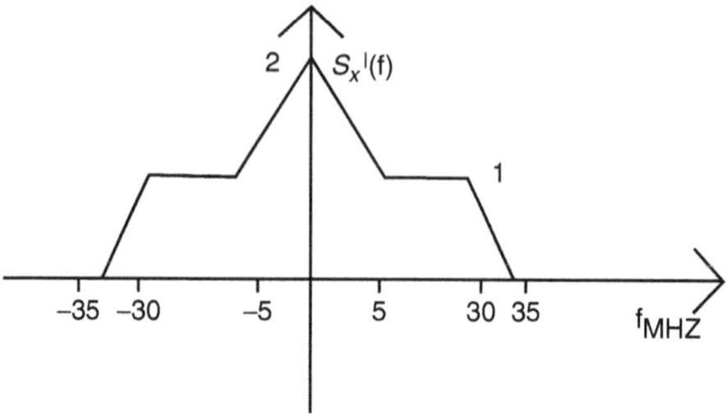

Fig. 3.12 Spectral density of the X_t^I which is same as that of the spectral density X_t^Q

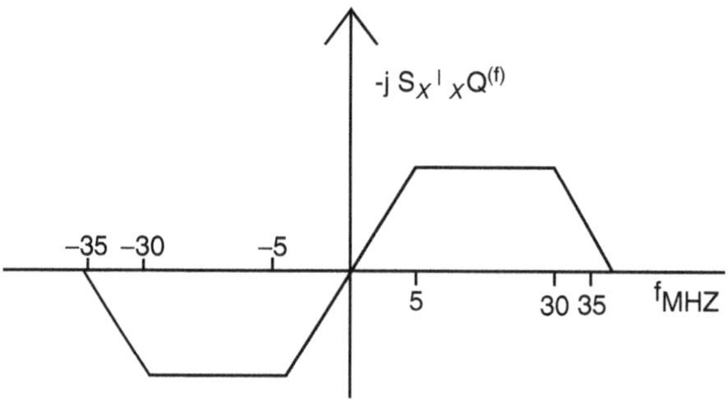

Fig. 3.13 Spectral density $-\mathrm{j}S_{X^I X^Q}(f)$

Chapter 4
Linear Algebra

4.1 Vector Space

1. The set of vectors forms the vector space 'V' over the Field 'F', if the vectors belonging to that set satisfy the following properties

   (a) If v1, v2 are the elements of the vector space 'V', then the vector defined by v1 + v2 must be the element of the vector space 'V'.
   (b) For some scalars {'α' 'β'} ∈ F.

$$(\alpha + \beta)v = \alpha v + \beta v.$$
$$(\alpha \beta)v = \alpha(\beta v).$$

   (c) There exists the identity scalar represented as '1' such that 1.v = v.1 = v.
   (d) There exists the additive identity vector represented as '0' such that for any vector 'v' ∈ vector space, v + 0 = 0 + v = v.
   (e) There exists the additive inverse vector for every vector 'v' ∈ vector space which is represented as '−v' such that v+(−v) = 0, where 0 is the additive identity vector which is the element of the vector space.
   (f) (v1 + v2)α = α v1 + α v2.
   (g) v1 + (v2 + v3) = (v1 + v2) + v3.
   (h) α v ∈ V where α ∈ Fv ∈ V.

2. 'W' is the subspace of the vector space 'V' (W ⊆ V) if the elements of the set 'W' is the subset of the vector space 'V' and satisfies all the properties mentioned in the fact 1.
3. If W1 ⊆ V and W2 ⊆ V, then W1 ∩ W2 ⊆ V. But W1 ∪ W2 ⊆ V only when W1 ⊆ W2 or W2 ⊆ W1.
4. The set of vectors {v1, v2, v3, v4...vn} are said to be linearly independent if α1 v1 + α2 v2 + α3 v3 + α4 v4 + ...αn vn = 0 for some scalars {α1, α2, α3...αn}, then α1 = α2 = α3 = ...αn = 0.
5. If the set of vectors {v1,v2,v3,...vn} are linearly dependent if any one of the vector in the set is represented as the linear combinations of other vectors.
6. The set of vectors {v1,v2,v3,...vn} forms the Generating set of the vector space 'V' if any vector in the vector space 'V' can be represented as the linear

E.S. Gopi, *Mathematical Summary for Digital Signal Processing Applications with Matlab*, DOI 10.1007/978-90-481-3747-3_4,

combinations of the vectors in the Generating set. Also note that the Generating set is the subset of the Vector space 'V'.

7. Generating set which are linearly independent is called the basis of the vector space 'V'.
8. If {v1,v2,v3,...vn} be the basis of the vector space 'V', then the set of (n + 1) vectors in the vector space is always dependent.
9. The set of minimum number of the vectors which forms the Generating set of the vector space 'V' is called as Minimal Generating Set.
10. The maximum number of independent vectors collected from the vector space "V is called maximal independent set.
11. If {u1,u2,u3,...un} is the minimal generating set, then the maximal independent set is {u1,u2,u3,...un}.
12. If {u1,u2,...un} is the maximal linear independent set then {u1,u2,u3,...un} is the basis of the vector space 'V'.
13. The number of elements of the set is called as the cardinal number of the set. The cardinal number of the Basis set is called dimension (dim) of the vector space 'V'.
14. If W is the subspace of V(W ⊆ V) then
    (a) the Basis of W ⊆ Basis of V.
    (b) Any basis of 'W' is extended to the basis of 'V'.
    (c) dim(W) ⊆ dim(V).
15. If W1 ⊆ V and W2 ⊆ V then
    (a) dim(W1 ∪ W2) = dim(W1) + dim(W2) – dim(W1 ∩ W2).
    (b) Let the basis for the subspace W1 ∩ W2 is {u1,u2,u3,...uk}.
    The basis for the subspace W1 is obtained by extending the basis of the subspace W1 ∩ W2 as {u1,u2,...uk,w1,w2,w3,...wm}. Similarly the basis for the subspace W2 is obtained by extending the basis of the subspace W1 ∩ W2 as {u1,u2,u3,...uk,v1,v2,v3,...vn}. Note that the dimension of the vector space W1 is k + m and the dimension of the vector space W2 is k + n.
    (c) The basis of the vector space 'V' is obtained as {u1,u2,...uk,w1,w2, w3, ...wm,v1,v2,v3,...vm}.

4.2 Linear Transformation

1. Let 'U' and 'V' be the vector spaces over the field F.A Map T: V → V is called linear transformation if
    (a) T(u + v) = T(u) + T(v)
    (b) T(α U) = α T(u)

    where u ∈ U and v ∈ V
    The Linear map is graphically represented as follows (Fig. 4.1).

Fig. 4.1 Linear transformation

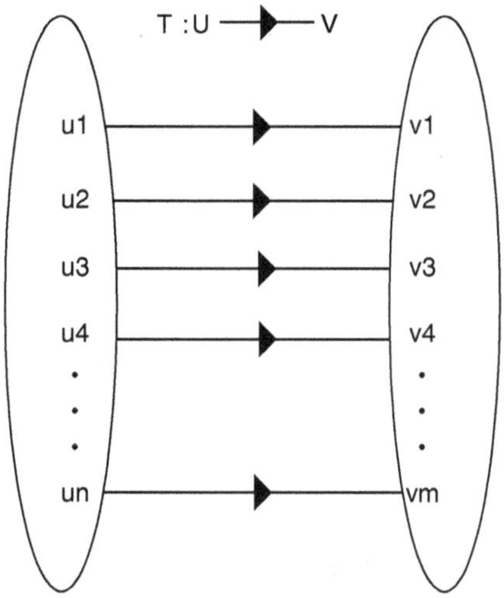

Fig. 4.2 Injective transformation

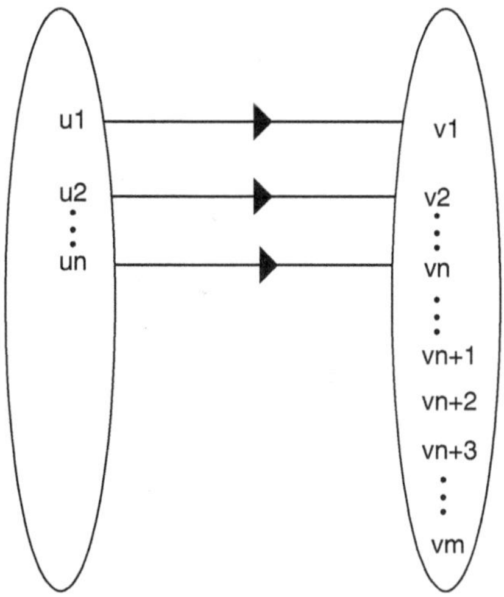

2. The linear transformations are broadly classified as (a) ONE-ONE (Injective) transformation (b) ONTO (Surjective) transformation (c) Isomorphic (Bijective) transformation.

   (a) Injective transformation (Fig. 4.2)
   A Linear map T: U $\rightarrow$ V is said to be injective if T(u1) $=$ T(u2)

   $$\Rightarrow \text{u1} = \text{u2}$$

   In this case ker(U) $= \{0\}$. Also note that dim(U) $<=$ dim(V)

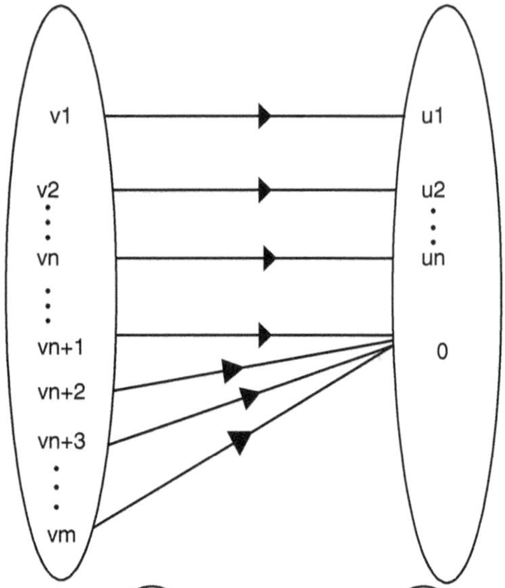

Fig. 4.3 Surjective transformation

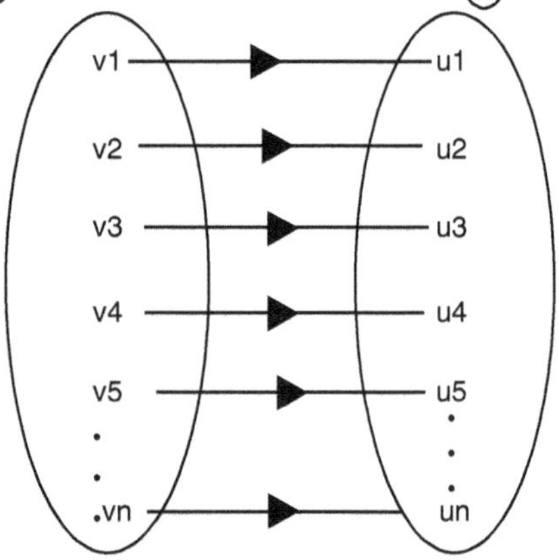

Fig. 4.4 Isomorphic transformation

(b) Surjective transformation (Fig. 4.3)
A Linear map T: V $\rightarrow$ U is said to be surjective if, for any w $\in$ W, there exists v $\in$ V, such that T(v) = w.
dim(V) <= dim(W). In this case Ker(V) $\neq$ {0}
(c) Isomorphic transformation (Fig. 4.4)

A Linear map T: V $\rightarrow$ U is said to be isomorphic, if the transformation is both Injective and Surjective in nature.

dim(V) = dim(W)
dim(Ker(V)) = 0
Ker(V) = {0}

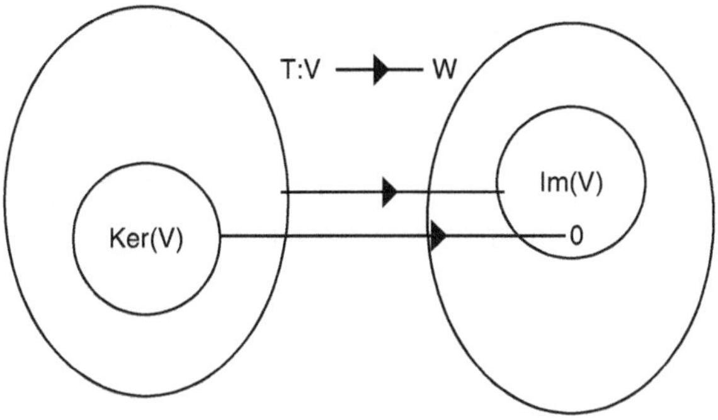

Fig. 4.5 Illustrations of the kernel and Image of the vector space

3. Let the linear transformation T: V → W is defined as the linear map from the vector space 'V' to 'W' (Fig. 4.5).

   (a) The image of the vector space 'V' represented as Im(V) is the set of vectors which are linearly mapped from all the vectors in the vector space V. Note that Im(V) is the subspace of the vector space 'W'.

   $$\text{Im(V)} = \{\text{Tu, for all u} \in \text{V}\}$$

   (b) The Kernel of the vector space 'V' represented as Ker(V) is the set of vectors in the vector space 'V' which are mapped to the zero vector (Additive identity) in the vector space 'W'. Note that the kernel of the vector space 'V' is the subspace of the vector space 'V'

   $$\text{Ker(V)} = \{\text{u such that Tu} = 0\}$$

4. Properties of the linear transformation

   (a) Two vector spaces are said to be isomorphic, if there exists the isomorphic transformation between them.
   (b) Isomorphic transformation between the vector space V and W exists only when dim(V) = dim(W).
   (c) The Linear transformation 'T' is one-one transformation if and only if the Transformation takes independent sets into other linearly independent sets.
   (d) Consider the Linear transformation T: U → W. The transformation 'T' is one-one transformation if there exists always pre-image (i.e.) if w ∈ W,t here exists u ∈ U such that Tu = w.
   (e) If (u1,u2,u3,. . .un) be the basis of the vector space U and (w1,w2,w3,. . .wn) be the basis of the vector space 'W', then their exists the unique transformation T, such that Tu1 = w1, Tu2 = w2 . . . Tun = wn.

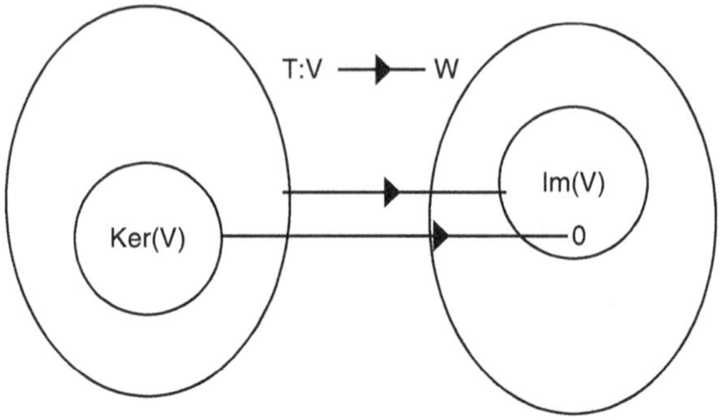

Fig. 4.6 Rank-Nullity theorem

5. **Rank-Nullity Theorem** (Fig. 4.6)
   Let the basis of ker(V) be {v1,v2,v3,...vn}. Extend this to the basis of V as

   $$\{\text{v1, v2, v3, \ldots vn, vn + 1, vn + 2, \ldots vm}\}.\text{Then the the set of vectors}$$
   $$\{\text{T(vn + 1), T(vn + 2), T(vn + 3)} \ldots \text{T(vm)}\} \text{ forms the basis of Im(V).}$$
   $$\Rightarrow \text{dim(V)} = \text{dim(Im(V))} + \text{dim(ker(V))}$$

6. Let the basis of the vector space 'V' be v1,v2,v3,...vm and the basis of the vector space 'W' be w1,w2,23,...wn. The set of all transformations from the vector space V to W is represented as L(V,W).L(V,W) is the vector space with dimension dim(V)dim(W). The basis of the vector space T(V,W) is represented as T11,T12,...T1n,T21,T22,T23,... T2n,T31, T32,..T3n,...Tm1,Tm2,Tm3,...Tmn which satisfies the following condition.

   $$\begin{aligned}\text{Tij (vk)} &= \text{wj if i} = \text{k}\\ &= 0, \text{otherwise}\end{aligned}$$

7. Let V be an n-dimensional vector space over F. Then the dual space V* is an n-dimensional vector space over F which consists of set of linear transformations and satisfies the following conditions (Fig. 4.7).

   (a) The Unique basis of the Dual space V* is the set of transformations {T1,T2,T3,...Tn}

   $$\begin{aligned}\text{such that Ti(vj)} &= \delta\text{ij, where } \delta\text{ij} = 1 \text{ if i} = \text{j}\\ &= 0, \text{otherwise}\end{aligned}$$

   (b) For any linear transformation T ∈ V*

   $$\text{T} = \sum_{i=1}^{n} T(vi)Ti$$

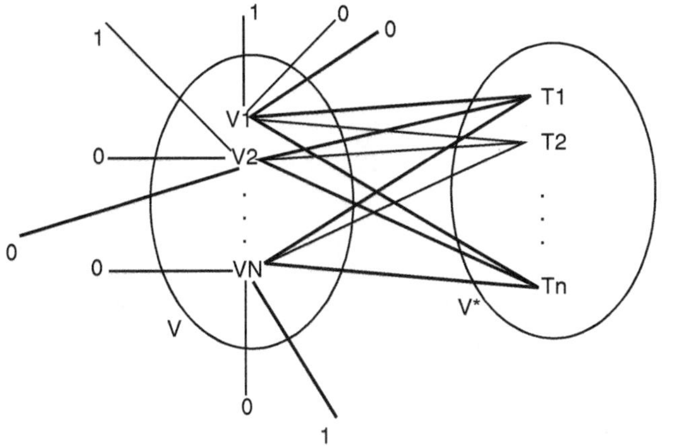

Fig. 4.7 Illustration of the dual space

(c) For any vector v ∈ V

$$V = \sum_{i=1}^{n} Ti(v)vi$$

Consider the arbitrary transformation T which is written as the linear combinations of the vectors T1,T2, T3,...Tn as $T = \alpha1^*T1 + \alpha2^*T2 + \alpha3^*T3 + \ldots \alpha n^*Tn$

$$\begin{aligned} T(v1) &= (\alpha1^*T1 + \alpha2^*T2 + \alpha3^*T3 + \ldots \alpha n^*Tn)v1 \\ &= \alpha1^*T1(v1) + \alpha2^*T2(v1) + \alpha3^*T3(v1) + \ldots \alpha n^*Tn(v1) \\ &= \alpha1 + 0 + 0 + \ldots + 0 \\ \Rightarrow T(v1) &= \alpha1 \end{aligned}$$

Similarly $T(v2) = \alpha2, T(v3) = \alpha3, \ldots, T(vn) = \alpha n$.

Consider the arbitrary vector V which is written as the linear combinations of the vectors v1,v2,v3,...Vn as $v = \beta1^*v1 + \beta2^*v2 + \beta3^*v3 + \ldots + \beta n^*vn$

$$\begin{aligned} Ti(v) &= Ti(\beta1^*v1 + \beta2^*v2 + \beta3^*v3 + \ldots \beta i^*vi + \ldots + \beta n^*vn) \\ &= 0 + 0 + 0 + \ldots \beta i + \ldots 0 \\ &= \beta i \end{aligned}$$

8. Set of all transformations acting on all the vectors in the subspace $W \subseteq V$ to get zeros are called Annihilator W^0 (Fig. 4.8) $\dim(w) + \dim(w^0) = \dim(V)$. Let the basis of w be {w1,w2,w3,...wk}. Extend the basis to the basis of V as {w1,w2,w3,...wk, wk +1,...wn}. $\{Tk+1, Tk+2, Tk+3, \ldots Tn\}$ is the basis of the w^0.
9. Any 'k' dimensional subspace is the intersection of (n − k) subspaces with dimension (n − 1) (Fig. 4.9).
10. Let ker(fi) = Ni. f is the linear combinations of f1,f2, f3,...fk if and only if $N1 \cap N2\ N3 \ldots \cap NK \subseteq N$.

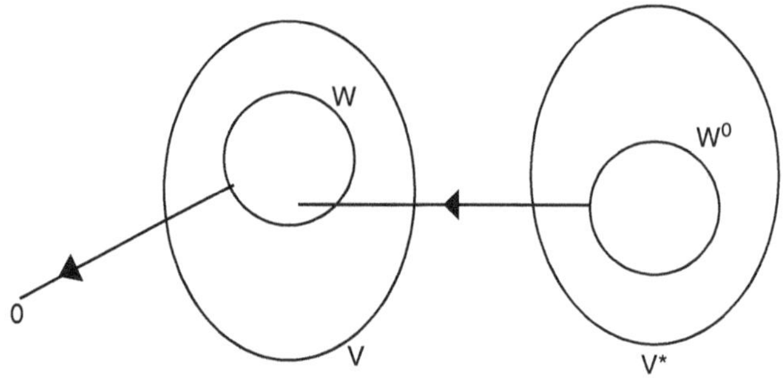

Fig. 4.8 Illustration of the annihilator

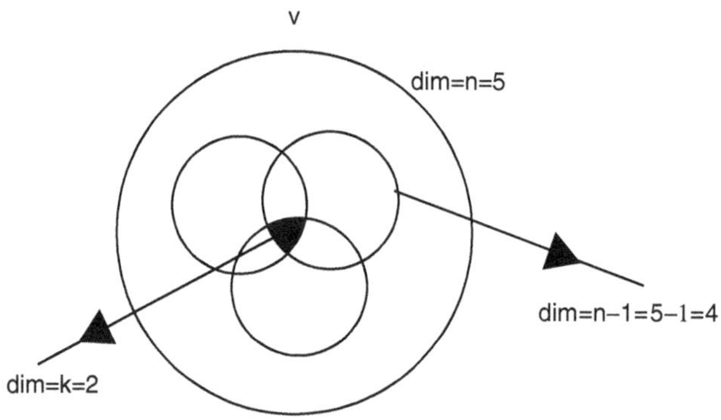

Fig. 4.9 Illustration for the property 9 of the linear transformation

11. Consider the transformation from the vector space V to W as T: V $\rightarrow$ W. Consider the dual space for the vector space V and W be represented as V* and W* respectively.

    There exists the transformation denoted by T^t : W* $\rightarrow$ V* such that the transformation T and T^t satisfies the following conditions.

    (a) $\mathrm{Im}(T^t) = \mathrm{Ker}(T)^0$ (Fig. 4.10)
    (b) $(\mathrm{Im}(T))^0 = \mathrm{Ker}(T^t)$ (Fig. 4.11)

4.3 Direct Sum

Let V1,V2,V3...Vn be the subspaces of the vector space V. The vector V is said to have direct sum representation (i.e.) V = V1 $\oplus$ V2 $\oplus$ V3 $\oplus$ ... Vn if the following conditions are satisfied

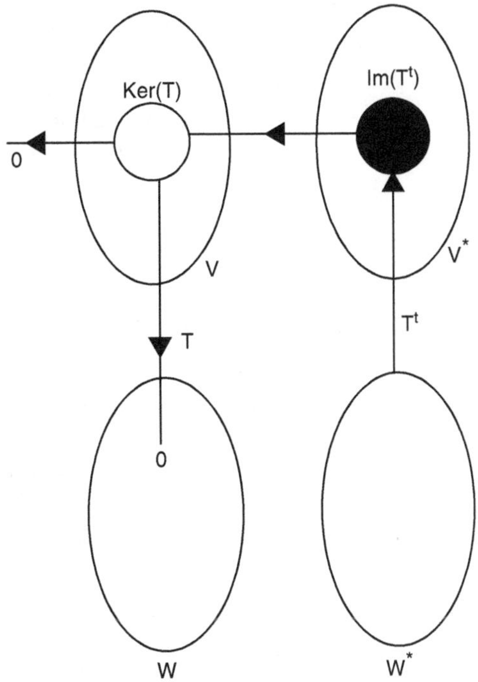

Fig. 4.10 Illustration of the property Im (Tt) = Ker(T)0

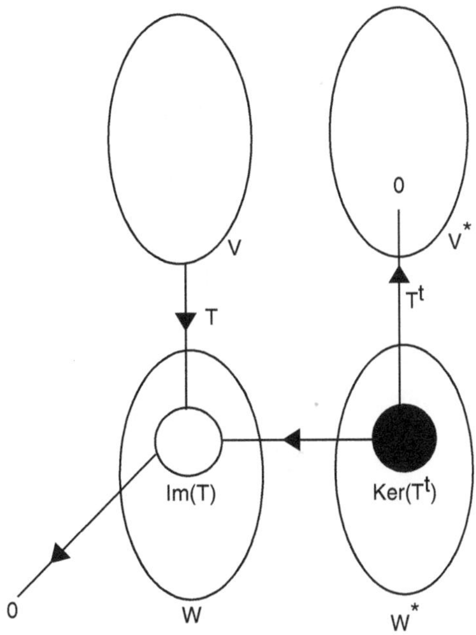

Fig. 4.11 Illustration of the property (Im(T)) 0 = Ker(Tt)

1. Any vector $v \in V$ is written uniquely as the summation of the vectors v1,v2,v3,...vn such that $v1 \in V1,\ v2 \in V2,\ v3 \in V3, \ldots vn \in Vn$.
2. Intersection of the vector space is the zero vector. $\cap Vi = \{\underline{0}\}$.
3. If $\underline{0}$ is represented as the summation of the vectors belongs to $V1, V2, \ldots Vn$, then the vectors collected from the vector spaces $V1, V2, \ldots Vn$ are $\underline{0}$.

$$\text{(i.e)}\ \underline{0} = \underline{0} + \underline{0} + \underline{0} + \cdots \underline{0}$$

4.4 Transformation Matrix

Consider the Isomorphic Linear transformation T from the vector space V to W.

$T : V \to W$. Let the basis of the vector space V and W are represented as $\{v1,\ v2,\ v3, \ldots vn\}$ and $\{w1, w2, w3, \ldots wm\}$ respectively. Linear transformation T acting on v1 which is represented as $T(v1) \in W$ can be represented as the linear combination of w1,w2,w3..wm as follows

$$
\begin{aligned}
&T(v1) = \alpha 11^{*}w1 + \alpha 12^{*}w2 + \alpha 13 * w3 + \cdots \alpha 1m^{*}wm\\
&T(v2) = \alpha 21^{*}w1 + \alpha 22^{*}w2 + \alpha 23 * w3 + \cdots \alpha 2m^{*}wm\\
&T(v3) = \alpha 31^{*}w1 + \alpha 32^{*}w2 + \alpha 33 * w3 + \cdots \alpha 3m^{*}wm\\
&\ldots\\
&T(vn) = \alpha n1^{*}w1 + \alpha n2^{*}w2 + \alpha n3 * w3 + \cdots \alpha nm^{*}wm
\end{aligned}
$$

Consider the arbitrary vector $v \in V$ which is represented as the linear combinations of the basis vectors $\{v1,\ v2,\ v3, \ldots vn\}$ as $\beta 1 * v1 + \beta 2 * v2 + \beta 3 * v3 + \ldots \beta n * vn$

$$
\begin{aligned}
T(v) &= T(\beta 1 * v1 + \beta 2 * v2 + \beta 3 * v3 + \cdots \beta n * vn)\\
\beta 1 &= {}^{*}T(v1) + \beta 2 * T(v2) + \beta 3 * T(v3) + \cdots \beta n * T(vn)\\
&= \beta 1^{*}(\alpha 11^{*}w1 + \alpha 12^{*}w2 + \alpha 13 * w3 + \cdots \alpha 1m^{*}wm)\\
&\quad +\beta 2^{*}(\alpha 21^{*}w1 + \alpha 22^{*}w2 + \alpha 23 * w3 \cdots \alpha 2m^{*}wm)\\
&\quad +\beta 3^{*}(\alpha 31^{*}w1 + \alpha 32^{*}w2 + \alpha 33 * w3 + \cdots \alpha 3m^{*}wm)\\
&\quad +\ldots\\
&\quad +\beta n^{*}(\alpha n1^{*}w1 + \alpha n2^{*}w2 + \alpha n3 * w3 + \cdots \alpha nm^{*}wm)\\
&= (\beta 1^{*}\alpha 11 + \beta 2^{*}\alpha 21 + \beta 3^{*}\alpha 31 + \ldots \beta n^{*}\alpha n1)^{*}w1\\
&\quad +(\beta 1^{*}\alpha 12 + \beta 2^{*}\alpha 22 + \beta 3^{*}\alpha 32 + \ldots \beta n^{*}\alpha n2)^{*}w2\\
&\quad +(\beta 1^{*}\alpha 13 + \beta 2^{*}\alpha 23 + \beta 3^{*}\alpha 33 + \ldots \beta n^{*}\alpha n3)^{*}w3\\
&\quad +\ldots\\
&\quad +(\beta 1^{*}\alpha 1m + \beta 2^{*}\alpha 2m + \beta 3^{*}\alpha m3m + \ldots \beta n^{*}\alpha nm)^{*}wm
\end{aligned}
$$

The transformed vector T(v) is represented as the linear combinations of the basis vector (w1, w2,...wm) with the coefficients as mentioned above.

The scalar coefficients used to represent the vector v using the basis of the vector space V is given as $(\beta 1, \beta 2, \beta 3, \beta 4, \ldots \beta n)$.

Similarly the scalar coefficients used to represent the vector T(v) using the basis of the vector space W is given as

$$((\beta 1 * \alpha 11 + \beta 2 * \alpha 21 + \beta 3 * \alpha 31 + \ldots \beta n * \alpha n1),\ (\beta 1 * \alpha 12 + \beta 2 * \alpha 22 + \beta 3 * \alpha 32 + \ldots \beta n * \alpha n2),\ (\beta 1 * \alpha 13 + \beta 2 * \alpha 23 + \beta 3 * \alpha 33 + \ldots \beta n * \alpha n3), \ldots (\beta 1 * \alpha 1m + \beta 2 * \alpha 2m + \beta 3 * \alpha m3m + \ldots \beta n * \alpha nm))$$

The scalar coefficients which are used to represent the vector v in the vector space V is related to the scalar coefficients used to represent the vector T(v) in the vector space W using the matrix as given below.

$$\begin{bmatrix} \alpha 11 & \alpha 21 & \ldots & \alpha n1 \\ \alpha 12 & \alpha 22 & \ldots & \alpha n2 \\ \ldots & \ldots & \ldots & \ldots \\ \alpha 1m & \alpha 2m & \ldots & \alpha nm \end{bmatrix}$$

The matrix mentioned above is called transformation matrix which is represented as

$$[T]_{B1}^{B2}$$

Note that the above transformation matrix is represented with respect to the basis B1 in the vector space V and the basis B2 in the vector space W.

Trick to obtain the transformation matrix for the linear transformation T: V → W

Let {v1, v2,...vn} and {w1, w2,...wn} be the basis of the vector space V and W respectively.

Obtain the transformation vector corresponding to the individual basis elements of the vector space V. Let it be {T(v1) T(v2) T(v3)...T(Vn)}.

Represent the transformed vector as the linear combinations of the basis vectors in the transformed domain W as follows

$$\mathrm{T(v1)} = \alpha 11^{*}\mathrm{w1} + \alpha 12^{*}\mathrm{w2} + \alpha 13 * w3 + \cdots \alpha 1m^{*}\mathrm{wm}$$
$$\mathrm{T(v2)} = \alpha 21^{*}\mathrm{w1} + \alpha 22^{*}\mathrm{w2} + \alpha 23 * w3 + \cdots \alpha 2m^{*}\mathrm{wm}$$
$$\mathrm{T(v3)} = \alpha 31^{*}\mathrm{w1} + \alpha 32^{*}\mathrm{w2} + \alpha 33 * w3 + \cdots \alpha 3m^{*}\mathrm{wm}$$
$$\ldots$$
$$\mathrm{T(vn)} = \alpha n1^{*}\mathrm{w1} + \alpha n2^{*}\mathrm{w2} + \alpha n3 * w3 + \cdots \alpha nm^{*}\mathrm{wm}$$

Form the matrix with first column filled with the scalar coefficients which are used to represent the transformed vector T(v1) (i.e.) {α11 α12 α13...α1m}. Similarly the second column is filled up with the scalar coefficients which are used to represent the transformed vector T(v2) (i.e.) {α21 α22 α23 ...α2m}.

4.5 Similar Matrices

Consider the Isomorphic Linear transformation T : V → V.

Consider the basis B1 = $\{v_1, v_2, v_3, \ldots v_n\}$ with respect to which the transformation matrix is represented as $\boldsymbol{T_{B1}^{B1}}$. Consider the basis B2 = $\{u_1, u_2, u_3, \ldots u_n\}$ with respect to which the transformation matrix is represented as $\boldsymbol{T_{B2}^{B2}}$.

The vector $u_1 \in$ V. Hence u_1 can be written as the linear combinations of the basis vectors B1 as mentioned below.

$$u1 = a_{11}v_1 + a_{21}v_2 + a_{31}v_3 + \ldots a_{n1}v_n$$

$$\ldots$$

$$\text{In general, } u_n = a_{1n}v_1 + a_{2n}v_2 + a_{3n}v_3 + \cdots + a_{nn}v_n$$

$$\text{Consider the vector } [x]_{B2} = \begin{matrix} x1 \\ x2 \\ x3 \\ \vdots \\ xn \end{matrix}$$

This indicates that the vector $[x]_{B2}$ is represented as the linear combinations of the basis B2 as follows.

$$x_{B2} = x_1u_1 + x_2u_2 + x_3u_3 + \cdots + x_nu_n$$

$$\begin{aligned} \Rightarrow x_{B2} = & x_1 * [a_{11}v_1 + a_{21}v_2 + a_{31}v_3 + \cdots + a_{n1}v_n] + \\ & x_2 * [a_{12}v_1 + a_{22}v_2 + a_{32}v_3 + \cdots + a_{n2}v_n] + \cdots \\ & x_n * [a_{1n}v_1 + a_{2n}v_2 + a_{3n}v_1 + \cdots + a_{nn}v_n \end{aligned}$$

$$\begin{aligned} \Rightarrow x_{B2} = & v_1 * [a_{11}x_1 + a_{12}x_2 + a_{13}x_3 + \cdots + a_{1n}x_n] + \\ & v_2 * [a_{21}x_1 + a_{22}x_2 + a_{23}x_3 + \cdots + a_{2n}x_n] + \cdots \\ & v_n * [a_{n1}x_1 + a_{n2}x_2 + a_{n3}x_3 + \cdots + a_{nn}x_n] + \end{aligned}$$

Thus the vector [x] with reference to the basis B1 is represented as follows

$$[x]_{B1} = \begin{matrix} a_{11}x_1 + a_{12}x_2 + a_{13}x_3 + \cdots + a_{1n}x_n \\ a_{21}x_1 + a_{22}x_2 + a_{23}x_3 + \cdots + a_{2n}x_n \\ a_{31}x_1 + a_{32}x_2 + a_{33}x_3 + \cdots + a_{3n}x_n \\ \vdots \\ a_{n1}x_1 + a_{n2}x_2 + a_{n3}x_3 + \cdots + a_{nn}x_n \end{matrix}$$

The vector $[x]_{B1}$ can be rewritten as

$$[x]_{B1} = \begin{vmatrix} a11 & a12 & a13 & . & . & a1n \\ a21 & a22 & a23. & . & . & a2n \\ a31 & a32 & a33 & . & . & a3n \\ . & . & . & . & . & . \\ . & . & . & . & . & . \\ an1 & an2 & an3 & . & . & ann \end{vmatrix} \begin{matrix} x1 \\ x2 \\ x3 \\ . \\ . \\ xn \end{matrix}$$

$$\Rightarrow [x]_{B1} = \begin{vmatrix} a11 & a12 & a13 & . & . & a1n \\ a21 & a22 & a23. & . & . & a2n \\ a31 & a32 & a33 & . & . & a3n \\ . & . & . & . & . & . \\ . & . & . & . & . & . \\ an1 & an2 & an3 & . & . & ann \end{vmatrix} [x]_{B2}$$

$$\Rightarrow [x]_{B1} = [M][x]_{B2}, \text{ where}$$

$$[\mathrm{M}] = \begin{matrix} a11 & a12 & a13 & . & . & a1n \\ a21 & a22 & a23. & . & . & a2n \\ a31 & a32 & a33 & . & . & a3n \\ . & . & . & . & . & . \\ . & . & . & . & . & . \\ an1 & an2 & an3 & . & . & ann \end{matrix}$$

Also as the matrix [M] is the invertible matrix

$$[x]_{B2} = [M]^{-1}\, [x]_{B1},$$

Consider the vector $[x]_{B2}$ is transformed to another vector using the transformation matrix $\boldsymbol{T}_{\boldsymbol{B2}}^{\boldsymbol{B2}}$ as $\left[\boldsymbol{T}_{\boldsymbol{B2}}^{\boldsymbol{B2}}\, [x]_{B2}\right]_{B2}$ with reference to the basis B2.

Consider the vector $[x]_{B2}$ is represented with respect to the basis B1 is given as

$$[[M][x]_{B2}]_{B1}$$

The above mentioned vector with respect to the basis B1 is transformed using the transformation matrix $\boldsymbol{T}_{\boldsymbol{B1}}^{\boldsymbol{B1}}$ as

$$\left\{\left[\boldsymbol{T}_{\boldsymbol{B1}}^{\boldsymbol{B1}}\right] [[M][x]_{B2}]_{B1}\right\} \mathrm{B1}$$

Note that the above vector is with respect to the basis B1. The obtained vector is represented with respect to the basis B2 as follows

$$\left\{M^{-1}\, \left\{\left[\boldsymbol{T}_{\boldsymbol{B1}}^{\boldsymbol{B1}}\right]\, [[M]\, [x]_{B2}]_{B1}\right\}_{\mathrm{B1}}\right\}_{\mathrm{B2}} \tag{4.1}$$

This must be same as the transformed vector of the vector $[x]_{B2}$ obtained using the transformation matrix $\boldsymbol{T}_{B2}^{B2}$ which is represented as

$$\text{as } \left[\boldsymbol{T}_{B2}^{B2}[x]_{B2}\right]_{B2} \tag{4.2}$$

Comparing both the Eqs. (1) and (2)

$$\text{We get } \boldsymbol{T}_{B2}^{B2} = [M]^{-1}\left[\boldsymbol{T}_{B1}^{B1}\right][M]$$

The matrices $\boldsymbol{T}_{B2}^{B2}$ and $\left[\boldsymbol{T}_{B1}^{B1}\right]$ are called as similar matrices.

In general, the two matrices A and B are said to be similar matrices, if there exists the invertible matrix P such that B $= [P]^{-1}\,[A]\,[P]$.

4.6 Structure Theorem

The transformation matrix for the particular linear transformation can be written in different form corresponding to the different basis vectors. Structure theorem deals with the technique for obtaining the simplest transformation matrix like diagonal matrix, upper triangular matrix etc. such that computation of the transformation becomes simpler and faster.

1. Consider the one-one transformation T : V $\rightarrow$ W with dim (V) $=$ n and dim (W) $=$ k, n $<$ k Let {v1, v2, v3, . . .vn} be the basis of the vector space V. {T(v1), T(v2), T(v3). . .T(vn)} exists in the vector space W. They are linearly independent. Extend the set {T(v1), T(v2), T(v3). . .T(vn), wn + 1, wn + 2, . . .wk}.

   The transformation matrix with respect to the above mentioned basis will look like below.

$$\begin{array}{ccccccccc}
1 & 0 & 0 & . & 0 & a1,n+1 & . & a1,k \\
0 & 1 & 0 & . & 0 & a2,n+1 & . & a2,k \\
0 & 0 & 1 & . & 0 & a3,n+1 & . & a3,k \\
0 & 0 & 0 & . & 0 & a4,n+1 & . & a4,k \\
. & . & . & . & . & . & . & . \\
. & . & . & . & . & . & . & . \\
. & . & . & . & . & . & . & . \\
0 & 0 & 0 & 0 & 1 & an,n+1 & . & an,k
\end{array}$$

2. Consider the onto transformation T : V $\rightarrow$ W with dim (V) $=$ n and dim (W) $=$ k, n $>$ k. Let {v1, v2, v3, . . .vn} be the basis of the vector space V. {T(v1), T(v2), T(v3). . .T(vk)} exists in the vector space W. They are linearly independent. The transformation matrix associated with the above basis is given as

$$\begin{array}{cccccccc}
1 & 0 & 0 & . & 0 & 0 & . & 0 \\
0 & 1 & 0 & . & 0 & 0 & . & 0 \\
0 & 0 & 1 & . & 0 & 0 & . & 0 \\
0 & 0 & 0 & . & 0 & 0 & . & 0 \\
. & . & . & . & . & . & . & . \\
. & . & . & . & . & . & . & . \\
. & . & . & . & . & . & . & . \\
0 & 0 & 0 & 0 & 1 & 0 & . & 0
\end{array}$$

3. Consider the isomorphism transformation $T : V \rightarrow W$ with $\dim(V) = n$ and $\dim(W) = n$. Let $\{v1, v2, v3, \ldots vn\}$ be the basis of the vector space V. $\{T(v1), T(v2), T(v3) \ldots T(vk)\}$ exists in the vector space W. They are linearly independent. The transformation matrix associated with the above basis is given as

$$\begin{array}{ccccc}
1 & 0 & 0 & . & 0 \\
0 & 1 & 0 & . & 0 \\
0 & 0 & 1 & . & 0 \\
0 & 0 & 0 & . & 0 \\
. & . & . & . & . \\
0 & 0 & 0 & 0 & 1
\end{array}$$

   Note that the columns of the transformation matrix in all the three cases are obtained using the technique described in the section **Trick to obtain the transformation matrix for the linear transformation T: V → W**.
4. Consider the transformation matrix T which is not of simple form.
   (a) Form the polynomial by setting the equation $\det(T - \lambda I) = 0$. The polynomial thus obtained is called **characteristic polynomial of the transformation matrix**
   (b) The roots of the Characteristic polynomial is called **Eigen values**. Let it be $\lambda_1\ \lambda_2\ \lambda_3\ \lambda_4 \ldots \lambda_k$ Let us consider the characteristic polynomial for the transformation matrix T be $p(x) = (x - \lambda 1)^{n1}(x - \lambda 2)^{n2}(x - \lambda 3)^{n3}(x - \lambda 4)^{n4} \ldots (x - \lambda n)^{nk}$
   (c) If p(x) is the characteristic polynomial of the transformation matrix A (as mentioned above), then p(T) is the zero matrix, where addition and multiplication are performed in the usual matrix operation and the constant term 'c' in the polynomial p(A) is represented as **cI**, where **'I'** is the identity matrix. **This is called Cayley-Hamilton theorem (i.e.).**

$$p(T) = (T - \lambda 1\, I)^{n1}(T - \lambda 2\, I)^{n2}(T - \lambda 3\, I)^{n3}(T - \lambda 4\, I)^{n4} \ldots (T - \lambda n\, I)^{nk} = 0$$

   (d) But there can be the polynomial q(x) with lesser degree compared to degree of p(x) which satisfies the condition $q(T) = 0$. This polynomial is called **minimal polynomial of the transformation matrix 'T' (i.e.).**

$$q(T) = (T - \lambda 1\, I)^{m1}(T - \lambda 2\, I)^{m2}(T - \lambda 3\, I)^{m3}(T - \lambda 4\, I)^{m4} \ldots (T - \lambda n\, I)^{mk} = 0$$

   where $mk \leq nk \forall k$

(e) Consider the minimal polynomial of the transformation matrix 'T' is of the form in which the values of $m1 = m2 = m3 \ldots = mk = 1$.

$$q(T) = (T - \lambda 1\, I)^1 (T - \lambda 2\, I)^1 (T - \lambda 3\, I)^1 (T - \lambda 4\, I)^1 \ldots (T - \lambda n\, I)^1 = 0$$

Consider $(T - \lambda 1\, I)v = 0 \Rightarrow T(v) = \lambda 1 v$, the transformed vector T(v) is the scaled version of the vector v with scaling value λ1. The Vector satisfying the above form is called Eigen vector. The set of all vectors satisfying the above equation forms the space and are called **Eigen space**. They are represented as $V_{\lambda 1}$. The basis of such vector space satisfying the above mentioned conditions are called **Eigen basis** of the transformation matrix T corresponding to the Eigen value 'λ1'.

Consider the transformation matrix T: V → V. Vector space V is said to be **invariant space** if for any vector v ∈ V such that T(v) ∈ V.

Eigen values of the transformation matrix T are $\lambda 1, \lambda 2, \lambda 3, \ldots \lambda k$. The Eigen space corresponding to the Eigen values 'λ1', 'λ2', 'λ3',... 'λ1k are represented as $V_{\lambda 1}\ V_{\lambda 2}\ V_{\lambda 3} \ldots V_{\lambda k}$ respectively. They are individually the invariant subspace of the vector space V.

Suppose if the dim(V) = n, then $\dim(V_{\lambda 1)}) + \dim(V_{\lambda 2)}) + \dim(V_{\lambda 3)}) + \ldots \dim(V_{\Lambda k)}) = n$

Let The basis of the vector space $V_{\lambda 1}$ (i.e.) the Eigen vectors be {v11 v12 v13 ... v1i}. The basis of the vector space $V_{\Lambda 2}$ is given as {v21 v22 v23 ... v2j}. Similarly the basis of the vector space $V_{\Lambda k}$ is represented as {vk1 vk2 vk3 ... vkm}. Then the basis of the vector space V is given as {v11 v12 v13 ... v1i v21 v22 v23 ... v2j ... vk1 vk2 vk3 ... vkm}. This indicate that the vector space V is obtained as the direct sum of the vector spaces $V_{\lambda 1}\ V_{\lambda 2}\ V_{\lambda 3} \ldots V_{\lambda k}$. (i.e.)

$$\mathbf{V = V_{\lambda 1} \oplus V_{\lambda 2} \oplus V_{\lambda 3} \ldots \oplus V_{\lambda k}}$$
$$T(v11) = \lambda 1\ v11$$
$$T(v12) = \lambda 1\ v12$$
$$T(v13) = \lambda 1\ v13$$
$$\ldots$$
$$T(v1i) = \lambda 1\ v1i$$
$$T(v21) = \lambda 2\ v21$$
$$T(v22) = \lambda 2\ v22$$
$$T(v23) = \lambda 2\ v23$$
$$T(v2j) = \lambda 2\ v2j$$
$$\ldots$$
$$T(vk1) = \lambda k\ vk1$$
$$T(vk2) = \lambda k\ vk2$$
$$\ldots$$
$$T(vkm) = \lambda k\ vkm$$

The transformation matrix associated with the above mentioned basis is given as follows.

$$\begin{array}{cccccccccc}
\lambda 1 & 0 & . & 0 & 0 & 0 & . & . & 0 & 0 \\
0 & \lambda 1 & . & 0 & 0 & 0 & . & . & 0 & 0 \\
0 & 0 & . & . & . & . & . & . & 0 & 0 \\
0 & 0 & . & . & . & . & . & . & 0 & 0 \\
0 & 0 & . & . & . & . & . & . & 0 & 0 \\
0 & 0 & . & \lambda 1 & 0 & 0 & . & . & 0 & 0 \\
0 & 0 & . & 0 & \lambda 2 & 0 & . & . & 0 & 0 \\
0 & 0 & . & 0 & 0 & \lambda 2 & . & . & 0 & 0 \\
. & . & . & . & . & . & . & . & . & 0 \\
. & . & . & . & . & . & . & . & . & 0 \\
. & . & . & . & . & . & . & . & \lambda k & 0 \\
0 & 0 & 0 & 0 & 0 & 0 & 0 & 0 & 0 & \lambda k
\end{array}$$

The transformation matrix is the square matrix. Also the diagonal elements of the matrix is filled up with the Eigen values [λ1 λ1 λ1 ... λ1 .. λ2 λ2 λ2 ... λk λk λk]

(f) Consider the minimal polynomial of the transformation matrix 'T'

$$q(T) = (T - \lambda 1\ I)^{m1}(T - \lambda 2\ I)^{m2}(T - \lambda 3\ I)^{m3}(T - \lambda 4\ I)^{m4} \ldots (T - \lambda n\ I)^{mk} = 0$$

where $mk \leq nk\ \forall k$

Also consider the vector satisfying the condition $(T - \lambda 1\ I)^{m1}v = 0$ (i.e.) the vector 'v' $\in \ker((T - \lambda 1\ I)^{m1})$. The vector v is called as **Generalized Eigen vector**. The set of all the vectors satisfying the above condition is called **Generalized Eigen space**. The Generalized Eigen space corresponding to the Eigen value 'λ1' is represented as V(λ1).

Consider the arbitrary vector $v1_1 \in V(\lambda 1)$. By the definition of minimal polynomial

$$(T - \lambda 1\ I)\, v1_1 \neq 0$$

$$\Rightarrow (T - \lambda 1\ I)\, v1_1 = v1_2$$

$$\Rightarrow T(v1_1) = \lambda 1 v1_1 + v1_2$$

Also $(T - \lambda 1\ I)^2 v1_1 = (T - \lambda 1\ I)(T - \lambda 1\ I)v1_1 = v1_3$

$$\Rightarrow (T - \lambda 1\ I)v1_2 = v1_3$$

$$\Rightarrow T(v1_2) = \lambda 1 v1_2 + v1_3$$

Similarly

$$T(v1_3) = \lambda 1 v1_3 + v1_4$$

$$T(v1_4) = \lambda 1 v1_4 + v1_5$$

$$T(v1_5) = \lambda 1 v1_5 + v1_6$$

$$T(v1_6) = \lambda 1 v1_6 + v1_7$$

$$\ldots$$

$$\text{Also } (T - \lambda 1\ \mathrm{I})^{m1} v_1 = (\mathrm{T} - \lambda 1\ \mathrm{I})(\mathrm{T} - \lambda 1\ \mathrm{I})^{m1-1} v_1 = 0$$

$$\text{But } (T - \lambda 1\ \mathrm{I})^{m1-1} v_1 = \mathrm{v_{m1}}$$

$$\Rightarrow (\mathrm{T} - \lambda 1\ \mathrm{I})\ \mathrm{v_{m1}} = 0$$

$$\Rightarrow \mathrm{T(v_{m1})} = \lambda 1\ \mathrm{v_{m1}}$$

Thus the independent vectors $\{v1_{m2}, v1_{m2-1}, v1_{m2-2}, v1_{m2-3}, v1_{m2-4}, v1_{m2-5} \ldots v1_1\}$ forms the basis for the vector space $V(\lambda 1)$. They are called **Jordon basis**
Similarly $\{v2_{m2}, v2_{m2-1}, v2_{m2-2}, v2_{m2-3}, v2_{m2-4}, v2_{m2-5} \ldots v2_1\}$ forms the basis for the vector space $V(\lambda 2)$ satisfying the similar conditions as described above.
In the same fashion $\{vk_{m2}, vk_{m2-1}, vk_{m2-2}, vk_{m2-3}, vk_{m2-4}, vk_{m2-5} \ldots vk_1\}$ forms the basis for the vector space $V(\lambda k)$ satisfying the similar conditions as described above.
Thus the set of independent vectors

$$\{v1_{m2}, v1_{m2-1}, v1_{m2-2} v1_{m2-3}, v1_{m2-4}, v1_{m2-5} \ldots v1_1, v2_{m2}, v2_{m2-1}, v2_{m2-2}, v2_{m2-3}, v2_{m2-4}, v2_{m2-5} \ldots v2_1, \ldots vk_{m2}, vk_{m2-1},$$

$vk_{m2-2}, vk_{m2-3}, vk_{m2-4}, vk_{m2-5} \ldots vk_1\}$ forms the basis of the vector space V.
Suppose the dim(V) = n, then dim(V(λ1)) + dim(V(λ2)) + dim(V(λ3)) + ...dim(V(λK)) = n
This indicate that the vector space V is obtained as the direct sum of the vector spaces $V(\lambda 1), V(\lambda 2), V(\lambda 3), ..V(\lambda k)$ (i.e.) $\mathbf{V = V(\lambda 1) \oplus V(\lambda 2) \oplus V(\lambda 3) \ldots \oplus V(\lambda k)}$
The transformation matrix with respect to the above mentioned basis is given as

$$\begin{array}{ccccccccccccccc}
\lambda 1 & 1 & 0 & 0 & . & 0 & 0 & 0 & . & . & . & 0 & 0 & 0 & 0 \\
0 & \lambda 1 & 1 & 0 & . & 0 & 0 & 0 & . & . & . & 0 & 0 & 0 & 0 \\
0 & 0 & \lambda 1 & 1 & . & 0 & 0 & 0 & . & . & . & 0 & 0 & 0 & 0 \\
0 & 0 & 0 & \lambda 1 & . & 0 & 0 & 0 & . & . & . & 0 & 0 & 0 & 0 \\
0 & 0 & 0 & 0 & . & 0 & 0 & 0 & . & . & . & 0 & 0 & 0 & 0 \\
. & . & . & . & . & . & . & . & . & . & . & 0 & 0 & 0 & 0 \\
. & . & . & . & . & . & . & . & . & . & . & 0 & 0 & 0 & 0 \\
. & . & . & . & . & . & . & . & . & . & . & 0 & 0 & 0 & 0
\end{array}$$

$$\begin{array}{ccccccccccccccc}
0 & 0 & 0 & 0 & . & \lambda 2 & 1 & 0 & . & . & . & 0 & 0 & 0 & 0 \\
0 & 0 & 0 & 0 & . & 0 & \lambda 2 & 1 & . & . & . & . & . & . & 0 \\
0 & 0 & 0 & 0 & . & 0 & 0 & \lambda 2 & . & . & . & . & . & . & 0 \\
. & . & . & . & . & . & . & . & . & . & . & . & . & . & 0 \\
. & . & . & . & . & . & . & . & . & . & . & \lambda k & 1 & 0 & 0 \\
. & . & . & . & . & . & . & . & . & . & . & 0 & \lambda k & 1 & 0 \\
0 & 0 & 0 & 0 & . & 0 & 0 & 0 & . & . & . & 0 & 0 & \lambda k & 1 \\
0 & 0 & 0 & 0 & 0 & 0 & 0 & 0 & 0 & 0 & 0 & 0 & 0 & 0 & \lambda k
\end{array}$$

Note that the matrix is filled up with Eigen values in the diagonal elements. Also note that upper off diagonal elements are filled up with one or zero.

4.7 Properties of Eigen Space

Consider the transformation matrix T: V → V Suppose for any vector $v \in V_\lambda \subseteq V$ such that $Tv = \lambda v$ for some scalar value 'λ', then the vector space V_λ is called Eigen space. The scalar value 'λ' is called as Eigen value. The vector satisfying the above condition is called Generalized Eigen vector. But in practice the basis of the Generalized Eigen space are referred as Generalized Eigen vector.

1. Eigen vectors corresponding to distinct Eigen values are independent.
2. The transformation matrix A is Diagonalizable if there exists Eigen vectors associated with the matrix A forms the basis of the vector space V.
3. If the matrix A is diagonalizable, the characteristic polynomial of the matrix A is represented as $(x - \lambda 1)^{n1}(x - \lambda 2)^{n2}(x - \lambda 3)^{n3}(x - \lambda 4)^{n4} \ldots (x - \lambda n)^{nk}$, Where $\dim(V_{\lambda k}) = nk$.
4. The vector space $V = V_{\lambda 1} \oplus V_{\lambda 2} \oplus V_{\lambda 3} \ldots \oplus V_{\lambda k}$ (Fig. 4.12).
5. The minimal polynomial of the diagonalizable matrix will always have the values $n1 = n2 = n3 = \ldots nk = 1$.
6. If $\dim(V_{\lambda 1}) = m1 < n1$, then the transformation matrix is called as deficient matrix. The value 'm1' is called Geometric multiplicity and 'n1' is called as Algebraic multiplicity.
7. The kernel $((A - \lambda i)^n) = V_{\lambda i}$, where n is the dimension of the vector space 'V'.
8. Image $((A - \lambda i)^n) = \oplus_{j=1}^{k} V_{\lambda j}$, where n is the dimension of the vector space 'V' $j \neq i$.
9. Any vector 'v' ∈ 'V' can be uniquely written as $v1 + v2$, where $v1 \in \text{kernel}((A - \lambda i)^n)$ and $v2 \in \text{Image}((A - \lambda i)^n)$.
10. The vector space V can be written as direct sum as given below

$$\text{kernel}\ ((A - \lambda i)^n) \oplus \text{Image}\ ((A - \lambda i)^n)$$

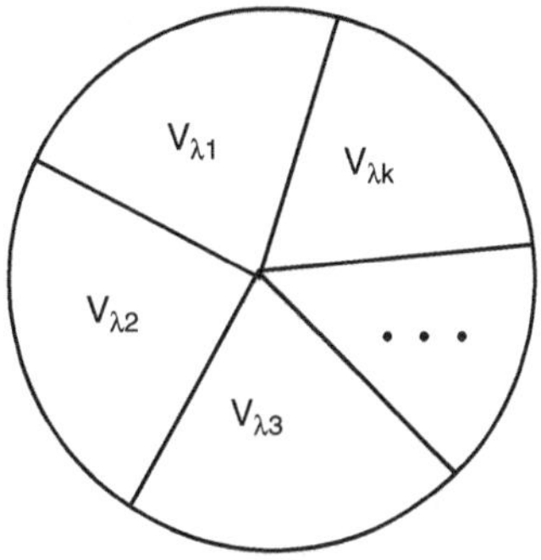

Fig. 4.12 Illustration of the direct sum of the Eigen space

4.8 Properties of Generalized Eigen Space

Consider the transformation matrix T: V → V Suppose for any vector v ∈ V(λ) ⊆ V such that $(T - \lambda I)^k v = 0$, where k is the minimal integer satisfying the condition, then the vector space V(λ) is called as Generalized Eigen space. The scalar value 'λ' is called as Eigen value. The vector satisfying the above condition is called Eigen vector. But in practice the basis of the Eigen space are referred as Eigen vectors.

1. $(A - \lambda I)$, $(A - \lambda I)^2$, $(A - \lambda I)^3, \ldots (A - \lambda I)^{k-1}$ are linearly independent.
2. Generalized Eigen vectors corresponding to distinct Eigen values are distinct.
3. If the minimal polynomial associated with the transformation matrix A is represented as

$$(x - \lambda 1)^{m1}(x - \lambda 2)^{m2}(x - \lambda 3)^{m3}(x - \lambda 4)^{m4} \ldots (x - \lambda n)^{mk}$$
$$\dim(V(\lambda 1)) = \dim(\text{kernel}((A - \lambda 1)^{m1})) = m1$$
$$\dim(V(\lambda 2)) = \dim(\text{kernel}((A - \lambda 2)^{m2})) = m2$$
$$\dim(V(\lambda 3)) = \dim(\text{kernel}((A - \lambda 3)^{m3})) = m3$$
$$\dim(V(\lambda 4)) = \dim(\text{kernel}((A - \lambda 4)^{m4})) = m4$$
$$\ldots$$
$$\dim(V(\lambda k)) = \dim(\text{kernel}((A - \lambda k)^{mk})) = mk$$

Also the vector space V = V(λ1) ⊕ V(λ2) ⊕ V(λ3) ... ⊕ V(λk) (Fig. 4.13)

4. The kernel $((A - \lambda i)^n)$ = V(λi), where n is the dimension of the vector space 'V'.
5. Image$((A - \lambda i)^n) = \oplus_{j=1}^{k}$ V(λj), where n is the dimension of the vector space 'V' and ≠ i.
6. Any vector 'v' ∈ 'V' can be uniquely written as v1 + v2, where v1 ∈ kernel $((A - \lambda i)^n)$ and v2 ∈ Image$((A - \lambda i)^n)$.
7. The vector space V can be written as direct sum as given below

$$\text{kernel}((A - \lambda i)^n) \oplus \text{Image}((A - \lambda i)^n)$$

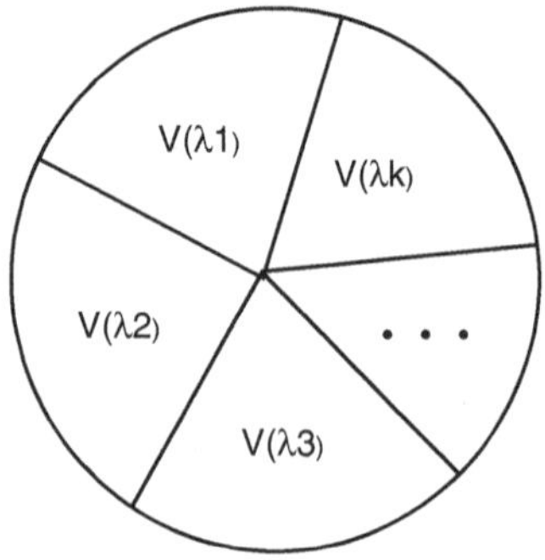

Fig. 4.13 Illustration of the direct sum of the generalized Eigen space

4.9 Nilpotent Transformation

1. The Matrix A is said to be Nilpotent, if ∃ the minimum integer k such that $A^k = 0$.
2. Consider the Linear transformation A: V → V such that A is the Nilpotent matrix is called Nilpotent Transformation.
3. If 0 is the only Eigen value of A, then A is Nilpotent.
4. '0' is the only Eigen value of the Nilpotent matrix.
5. Maximum possible value for k is 'n', where 'n' is the dim(A).
6. If A is Nilpotent and diagonalizable, then A = 0.
7. *If* $V_i = \ker(A^i),\ V_i \subset V_{i+1}$

$$\text{Let } v_i \in V_i = \ker(A^i)(i.e)A^i v_i = 0$$
$$A^{i+1}v_i = A(A^i v_i) = A(0) = 0$$
$$\Rightarrow v_i \ is\ the \ker(A^{i+1})$$
$$\Rightarrow V_i \subset V_{i+1}$$

Note that there can be the vector v_i which is the kernel of A^{i+1}, but not the kernel of A^i, but not the kernel of A^i (Fig. 4.14).

$$0 \subset V_1 \subset V_2 \subset V_3 \subset \cdots V_k = V$$

8. *If* $V_i = \ker(A^i), AV_i \subseteq V_{i-1}$
   Let $v_i \in V_i = \ker(A^i)(i.e.)A^i v_i = 0$

$$\Rightarrow A^{i-1}Av_i = 0$$
$$\Rightarrow Av_i \text{ is in the kernel of } V_{i-1}$$
$$\Rightarrow AV_i \subseteq V_{i-1}$$

9. A is Nilpotent, ∃ the matrix associated with A which is strictly upper triangular (Fig. 4.15).

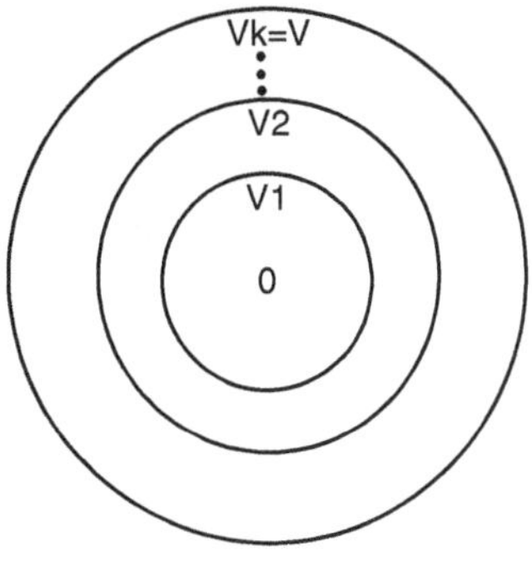

Fig. 4.14 Illustration of the Nilpotent property 7

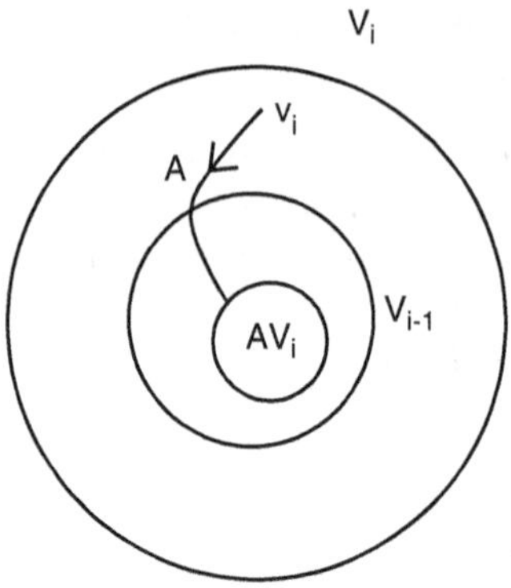

Fig. 4.15 Illustration of the Nilpotent property 9

Consider the basis of the V_1 as $\{v_{11}v_{12}v_{13}\ldots v_{1m1}\}$

Extend the basis of V_1 to the basis of V_2 as

$$\{v_{11}v_{12}v_{13}\ldots v_{1m1}\ v_{21}v_{22}v_{23}\ldots v_{2m2}\}$$

Similarly extend the basis of Vk − 1 to the basis of Vk as

$$\{v_{11}v_{12}v_{13}\ldots v_{1m1}\ v_{21}v_{22}v_{23}\ldots v_{2m2}\ldots v_{k1}v_{k2}v_{k3}\ldots v_{kmk}\}$$

The matrix corresponding to the above basis is strictly upper triangular matrix as shown below.

$$Av_{11} = 0$$
$$Av_{12} = 0$$
$$Av_{1m1} = 0$$

⇒ First m1 columns of the transformation matrix corresponding to the above basis is completely filled up with zeros

$$Av_{21} = ?$$
$$AV_2 \subseteq V_{2-1} = V_1$$
$$v_{21} \in V_2$$

⇒ v_{21} can be written as the linear combinations of the basis of V_1 $\{v_{11}v_{12}v_{13}\ldots v_{1m1}\}$. Let the coefficient elements are $\{\alpha_{11}\alpha_{12}\alpha_{13}\alpha_{14}\ldots\alpha_{1m1}\}$.

Similarly v_{2m2} can be written as the linear combinations of the basis of V_1 $\{v_{11}v_{12}v_{13}\ldots v_{1m1}\}$. Let the coefficient elements are represented as $\{\alpha_{m21}\alpha_{m22}\ \alpha_{m23}\alpha_{m24}\ldots\alpha_{m2m1}\}$.

Consider the representation of Av_{31}.

$$AV_3 \subseteq V_{3-1} = V_2$$
$$v_{31} \in V_3$$

⇒ v_{31} can be written as the linear combinations of the basis of V_2

$\{v_{11}v_{12}v_{13}\ldots v_{1m1}v_{21}v_{22}v_{23}\ldots v_{2m2}\}$. Let the coefficient elements are $\{\beta_{11}\beta_{12}\beta_{13}\beta_{14}\ldots\beta_{1m2}\}$

Similarly v_{3m3} can be written as the linear combinations of the basis of V_2

$$\{v_{11}v_{12}v_{13}\ldots v_{1m1}\ v_{21}v_{22}v_{23}\ \ldots v_{2m2}\}$$
$$\{\beta_{m31}\beta_{m32}\beta_{m33}\beta_{m34}\ldots\beta_{m3m2}\}$$

Similarly the Transformation acting on the other basis vectors are represented as the linear combinations of the basis vectors. Thus the transformation matrix corresponding to the above basis vectors is as shown below.

$$\begin{array}{cccccccccccc}
0 & . & 0 & \alpha_{11} & . & . & \alpha_{m21} & . & . & \beta_{11} & . & \beta_{m31} \\
0 & . & 0 & . & . & . & \alpha_{m22} & . & . & \beta_{12} & . & \beta_{m32} \\
0 & . & 0 & \alpha_{1m1} & . & . & . & . & . & \beta_{13} & . & \beta_{m33} \\
0 & . & 0 & 0 & . & . & . & . & . & \beta_{14} & . & \beta_{m34} \\
0 & . & 0 & 0 & 0 & . & . & . & . & \beta_{15} & . & . \\
0 & . & 0 & 0 & 0 & 0 & \alpha_{m2m} & . & . & . & . & . \\
0 & . & 0 & 0 & 0 & 0 & 0 & . & . & . & . & . \\
0 & . & 0 & 0 & 0 & 0 & 0 & 0 & . & . & . & . \\
0 & . & 0 & 0 & 0 & 0 & 0 & 0 & 0 & \beta_{1m2} & . & . \\
0 & . & 0 & 0 & 0 & 0 & 0 & 0 & 0 & 0 & . & . \\
0 & . & 0 & 0 & 0 & 0 & 0 & 0 & 0 & 0 & 0 & \beta_{m3m2} \\
0 & . & 0 & 0 & 0 & 0 & 0 & 0 & 0 & 0 & 0 & 0
\end{array}$$

4.10 Polynomial

Consider the polynomial P(x) is factored and written as the product of the polynomials as $p(x) = p1(x)p2(x)p3(x)p4(x)\ldots pk(x)$ and consider the linear transformation matrix A: V $\rightarrow$ V. Note that p(A) is matrix obtained using the polynomial p(x),where addition and multiplication are performed in the usual matrix operation and the constant term ‘c’ in the polynomial p(A) is represented as **cI**, where ‘**I**’ is the identity matrix.

1. kernel $(p(A)) = \oplus$ kernel(pi(A)).
2. The set of polynomials $q1(x)q2(x), q3(x)\ldots qk(x)$ are defined as follows:

$$q1(x) = p2(x)p3(x)\ldots pk(x)$$
$$q2(x) = p1(x)p3(x)\ldots pk(x)$$
$$\ldots$$
$$qk(x) = p1(x)p2(x)\ldots pk-1(x)$$

They are relatively prime.

3. Statement 2 implies there exists another set of vectors f1(x), f2(x), ... fk(x) such that f1(x)q1(x) + f2(x)q2(x) + ... fk(x)qk(x) = 1.
4. There exists the polynomial acting on some vector v ∈ ker(p(A)),gives the vector vi ∈ ker (pi(A)), for all i varies from 1 to k.
5. That polynomial is given as follows:
   q1(A)f1(A) is the polynomial when acted on the vector v ∈ ker (p(A)), gives the vector v1 ∈ ker (p1(A))
6. In general, qi(A)fi(A) is the polynomial when acted on the vector v ∈ ker(p(A)), gives the vector vi ∈ ker(pi(A)).

Note: The polynomial properties mentioned above can be compared with the minimal polynomial properties.

4.11 Inner Product Space

Consider the vector space V over the field F. Inner product over the vector space V is defined as the map from VXV to F satisfying the following axioms. It is represented as < V, V >. Note that VXV is the vector space with first element

1. < v, v >≠ 0 where v ∈ V
   If < v, v >= 0 then v = 0
2. < v, w + u >=< v, w > + < v, u > where u, v ∈ V
3. < cv, w > c < v, w > where c is the scalar constant.
4. < v, w >= $\overline{<w, v>}$, where $\overline{<w, v>}$, is called as conjugate

$$\Rightarrow < v, cw >= \bar{c} < w, v >$$

The vector space V with the with the defined inner product forms the Inner product space.

Examples for the inner product space

1. Consider the vector space $\mathcal{R}^2$. Consider two arbitrary vectors v1 = (a1, a2) ∈ $\mathcal{R}^2$, and v2 = (b1, b2) ∈ $\mathcal{R}^2$, then the inner product defined in the vector space V as < v1, v2 >= $a1\overline{b1} + a2\overline{b2}$. Thus the vector space $\mathcal{R}^2$ with the above defined inner product forms the inner product space.
2. Consider the vector space $\mathcal{R}^{nxn}$. Consider the arbitrary vector B ∈ $\mathcal{R}^{nxn}$, the inner product is defined as < A, B > trace(AB^*), B^*, is the conjugate of the matrix B. Thus the vector space $\mathcal{R}^{nxn}$ for the above defined inner product forms the inner product space.
3. Consider the vector space consists of the set of all complex valued functions defined for the interval (0,1] with the inner product defined as follows:

$$< \text{f, g} >= \int_0^1 f(t)\overline{g(t)dt} \text{ forms the inner product space.}$$

Formation of Inner product space for the vector space V
Consider the n-dimensional vectors space V. Let the basis of the vector basis be represented as {v1, v2, v3...vn}.

There exists the transformation A: V $\rightarrow$ V such that Av1 $=$ e1, Av2 $=$ e2, ... Avn $=$ en, where 'ei' is the vector with n elements completely filled up with zeros except ith element which is filled up with 1.e1, e2, e3, ... en are called as standard basis of the vector space $\mathcal{R}^n$.

The Inner product of the vectors u, v $\in$ V represented as $< v, u >$ is defined as follows.

$$\text{Let } v = \alpha 1^* v1 + \alpha 2^* v2 + \alpha 3^* v3 + \ldots \alpha n^* vn$$
$$u = \beta 1^* v1 + \beta 2^* v2 + \beta 3^* v3 + \ldots \beta n^* vn$$
$$< v,\ u >= \alpha 1\overline{\beta 1} + \alpha 2\overline{\beta 2} + \alpha 3\overline{\beta 3} + \alpha 4\overline{\beta 4} \ldots + \alpha n\overline{\beta n}$$

Norm of the vector v is denoted as follows $\|v\| = \sqrt{< v, v >}$

Properties of the norm

1. $\|c, u\| = |c|\|u\|$
2. $\|u\| > 0$
3. Cauchy-Schwarz inequality $| < u, v > | \leq \|u\|\ \|v\|$
4. $\|u + v\| \leq \|u\| + \|v\|$
5. $| < v, u > | \geq \text{Re}(< v, u >)$

4.12 Orthogonal Basis

1. Consider the inner product vector space V. Let u, v $\in$ V. The vector 'u' is orthogonal to the vector 'v' if the inner product $< u, v >= 0$.
2. The set of vectors {v1, v2, ... vk} are orthogonal set if $< vi,\ vj >= 0$, where $i \neq j$ and i, j = 1, 2, ..k.
3. The set of orthogonal vectors are always linearly independent.
4. If the set of Orthogonal vectors {v1, v2, ... vk} with $\|vi\| = 1$, $i = 1..k$ are called as orthonormal vectors.
5. If the set of orthogonal vectors are {v1, v2, ... vk}, then the set of orthonormal vectors are obtained as $\{\frac{v1}{\|v1\|}, \frac{v2}{\|v2\|}, \ldots \frac{vk}{\|vk\|}\}$.
6. If the set of orthogonal vectors forms the basis of the vector space V, they are called orthogonal basis.
7. If {v1, v2, v3, ... vn} forms the basis of the vector space V, then there exists the set of vectors {u1, u2, u3, ... un} which is the orthonormal set which forms the basis of the vector space 'V', which are obtained using Gram-Schmidt orthogonalization procedure as given below.

$$v1' = \frac{v1}{\|v1\|} = u1$$

$$
\begin{aligned}
\text{v2'} &= \text{v2} - < \text{v2}, \text{u1} > \text{u1} \\
\text{u2} &= \frac{v2'}{\|v2'\|} \\
\text{v3'} &= \text{v3} - < \text{v3}, \text{u1} > \text{u1} - < \text{v3}, \text{u2} > \text{u2} \\
\text{u3} &= \frac{v3'}{\|v3'\|}
\end{aligned}
$$

and so on.

8. Set of all vectors in the vector space V which are orthogonal to the set of vectors of the subspace $S \subset V$ is the vector space represented as $\mathbf{S}^{\perp}$.
9. If the dimension of the vector space S is k, then the dimension of the vector space $\mathbf{S}^{\perp}$ is n − k, where n is the dimension of the vector space 'V' (Fig. 4.16).
10. If the basis of the space S is $\{w_1, w_2, \ldots w_{k-1}, w_k\}$ and the basis of the vector space $\mathbf{S}^{\perp}$ is $\{w_{k+1},\ w_{k+2}, \ldots w_{n-1},\ w_n\}$, then the basis of the vector space V is given as $\{w_1, w_2, \ldots w_{k-1}, w_k,\ w_{k+1}, w_{k+2}, \ldots w_{n-1},\ w_n\}$.

$$\text{(i.e.) } V = S \oplus \mathbf{S}^{\perp}$$

11. The basis are orthogonal to each other between the space S and $\mathbf{S}^{\perp}$.
12. Any vector $v \in V$ can be uniquely written as v1 + v2, where $v1 \in S$ and $v2 \in \mathbf{S}^{\perp}$
13. Any vector $v \in V$ can be written as the linear combinations of the
14. Orthonormal basis vectors {v1,v2,v3,... vn} as $\alpha_1 v1 + \alpha_2 v2 + \alpha_3 v3 + \ldots \alpha_n vn$, where

$$
\begin{aligned}
&\alpha_1 \text{ is computed as } < \text{v}, \text{v1} > . \\
&\text{Similarly } \alpha_2 =< v, v2 > \\
&\alpha_3 =< \text{v}, \text{v3} > \\
&\alpha_4 =< \text{v}, \text{v4} > \\
&\text{In general } \alpha_k =< v, vk >
\end{aligned}
$$

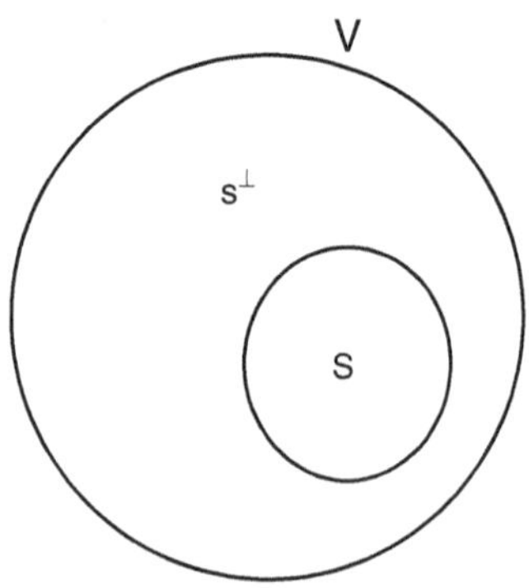

Fig. 4.16 Illustration of the vector space S and its orthogonal complement $\mathbf{S}^{\perp}$

4.13 Riegtz Representation

Consider the Linear transformation T: V $\rightarrow$ F, where the vector space V with dimension 'n'. There exists unique vector y $\in$ V such that T(v) $=<$ v, y $>$

Technique to obtain the vector y $\in$ V

1. Find the kernel(T) = W
2. Find the $W^{\perp}$.
3. Find any orthonormal vector u $\in$ $W^{\perp}$
4. Compute $\overline{T(u)}$
5. The vector y = $\overline{\overline{T(u)}}$ u

Example 4.1. Consider the transformation

$$
\begin{aligned}
&A : R^n \rightarrow R \\
&A(R^n) = A([x1\, x2\, x3\, \ldots xn]) \\
&= \alpha_1 x_1 + \alpha_2 x_2 + \alpha_3 x_3 + \ldots + \alpha_n x_n \\
&\Rightarrow < (x_1 x_2 x_3\, \ldots x_n), (\alpha_1 \alpha_2\, \ldots\, \alpha_n) > \\
&= \alpha_1 x_1 + \alpha_2 x_2 + \alpha_3 x_3 + \ldots + \alpha_n x_n
\end{aligned}
$$

Where y = $(\alpha_1 \alpha_2 \ldots \alpha_n)$ is the unique vector $\in \mathcal{R}^n$.

Chapter 5
Optimization

5.1 Constrained Optimization

Consider the function $f(x, y) = 2x^2 + 4y + 3$. The requirement is to find out the optimal values for 'x' and 'y' such that f(x, y) is minimized. Also it has to satisfy the constraint that $g(x, y) = x + y + 3 = 0$. Let the local extremum (maximum or minimum) point be (x_0, y_o).

The curve $x + y + 3 = 0$ can be viewed as the set of points $(\alpha, 3-\alpha)$, $\forall\, \alpha \in \mathbb{R}$. This is known as parametric representation of the curve. The tangent vector at the point $(x_0,\ y_o)$ points towards the direction $\left(\frac{d\alpha}{d\alpha}, \frac{d(3-\alpha)}{d\alpha}\right) = [1\ -1]$. Also the gradient vector of the equation $g(x, y)$ is obtained as $\begin{bmatrix} \frac{\partial(x+y+3)}{\partial x} \\ \frac{\partial(x+y+3)}{\partial y} \end{bmatrix} = \begin{bmatrix} 1 \\ 1 \end{bmatrix}$ Note that the gradient vector and the tangent vector is always orthogonal to each other (see Fig. 5.1).

The solution of the above optimization problem lies on the intersection of the curve $g(x, y)$ and $f(x, y)$ So consider the points on the curve $g(x, y)$ represented in the parametric form as $g(x(\alpha), y(\alpha)) = g1(\alpha)$. Let the optimal point (x_0, y_o) be represented in terms of the variable 'α' as α_0. Note that the point lies on the curve $f(x, y)$. The function $f(x, y)$ can be represented as the function of 'α' for the points of intersection of the curves $g(x, y)$ and $f(x, y)$ as $f1(\alpha)$

Using Taylor series,

$$f1(\alpha + \alpha_0) = f1(\alpha_0) + \alpha\frac{df1(\alpha_0)}{d\alpha} + \frac{\alpha^2}{2!}\frac{d^2 f1(\alpha_0)}{d\alpha^2} + \dots$$

$$\Rightarrow f1(\alpha + \alpha_0) - f1(\alpha_0) = \alpha\frac{df1(\alpha_0)}{d\alpha} + \frac{\alpha^2}{2!}\frac{d^2 f1(\alpha_0)}{d\alpha^2} + \dots$$

$f1(\alpha_0)$ corresponds to local extremum and hence $f1(\alpha + \alpha_0) - f1(\alpha_0)$ must be greater than 0 if the point α_0 is local minima, or it must be lesser than 0 if the point α_0 is local maxima. Also α_0 can be either positive or negative value. Thus the condition that α_0 belongs to the local extremum is $\frac{df1(\alpha_0)}{d\alpha} = 0$. Also note that if

E.S. Gopi, *Mathematical Summary for Digital Signal Processing Applications with Matlab*, DOI 10.1007/978-90-481-3747-3_5,

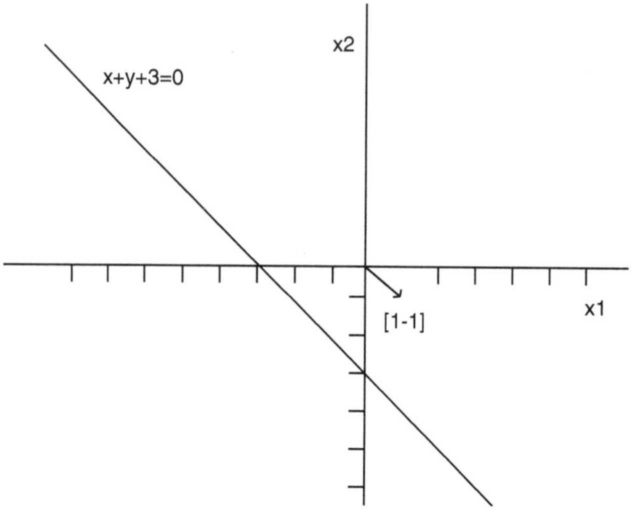

Fig. 5.1 Illustration of the property that the gradient vector is orthogonal to the tangent vector

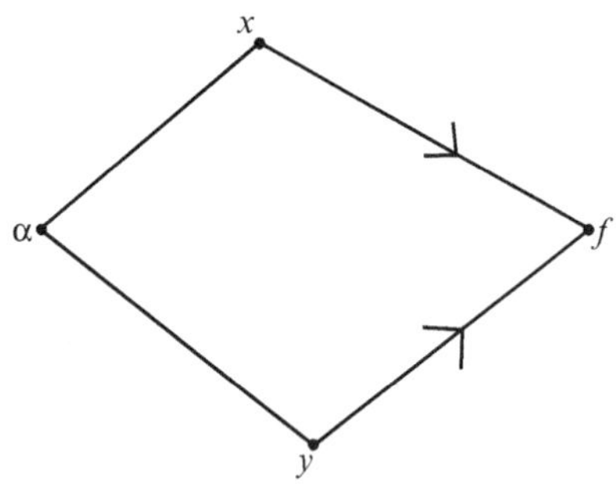

Fig. 5.2 Illustration of the functional dependencies

the extremum is maxima, then $\frac{d^2 f1(\alpha_0)}{d\alpha^2}$ must be negative and if the extremum is minima, then $\frac{d^2 f1(\alpha_0)}{d\alpha^2}$ is positive.

Thus the condition that the point α_0 belongs to extremum is $\frac{df1(\alpha_0)}{d\alpha} = 0$ (Fig. 5.2)

$$f1(\alpha_0) = f(x(\alpha_0), y(\alpha_0))$$

$$\Rightarrow \frac{df1(\alpha_0)}{d\alpha} = \frac{df(x(\alpha_0), y(\alpha_0))}{d\alpha} = \frac{df(x(\alpha), y(\alpha))}{d\alpha} at\ \alpha = \alpha_0$$

Using the illustration of functional dependencies as shown above, we get the following

$$\frac{df(x(\alpha), y(\alpha))}{d\alpha}(at\ \alpha = \alpha_0) = \frac{df}{dx} * \frac{dx}{d\alpha} + \frac{df}{dy} * \frac{dy}{d\alpha} = 0$$

$$\Rightarrow \begin{bmatrix} \frac{df}{dx} & \frac{df}{dy} \end{bmatrix} \begin{bmatrix} \frac{dx}{d\alpha} \\ \frac{dy}{d\alpha} \end{bmatrix} (at\ \alpha = \alpha_0) = 0$$

The vector $\begin{bmatrix} \frac{df}{dx} \\ \frac{df}{dy} \end{bmatrix}$ is the gradient vector of the curve $f(x, y)$ at the point $\alpha = \alpha_0$.

Also $\begin{bmatrix} \frac{dx}{d\alpha} \\ \frac{dy}{d\alpha} \end{bmatrix}$ at $\alpha = \alpha_0$ is the direction of the tangent vector on the point $\alpha = \alpha_0$ of the curve $g(x, y)$. [This is because of the fact that the set of points $((x(\alpha), y(\alpha))$ are the set of parametric representation of the points on the curve $g(x, y)$. Thus from the above statement, the gradient vector of the curve $f(x, y)$ at the point $(x(\alpha_0), y(\alpha_0))$ is orthogonal to the direction of the tangent vector drawn at the point $(x(\alpha_0), y(\alpha_0))$ on the curve $g(x, y)$.

We have already shown that the gradient vector of the curve $g(x, y)$ at the point $(x(\alpha_0), y(\alpha_0))$ and the direction of the tangent vector of the curve $g(x, y)$ at the point $(x(\alpha_0), y(\alpha_0))$ are orthogonal to each other.

This implies that the gradient vector of the curve $f(x, y)$ and the gradient vector of the curve $g(x, y)$ are parallel to each other and hence for some scalar 'λ', which is known as Lagrange multiplier,

Gradient vector of the curve $f(x, y)$ + λ* Gradient vector of the curve $g(x, y) = 0$.

Gradient vector of the curve $f(x, y)$ is represented as $\nabla f = \begin{bmatrix} \frac{df}{dx} \\ \frac{df}{dy} \end{bmatrix}$. Similarly the gradient vector of the curve $g(x, y)$ is represented as $\nabla g = \begin{bmatrix} \frac{dg}{dx} \\ \frac{dg}{dy} \end{bmatrix}$.

$$\Rightarrow \nabla f + \lambda \nabla g = 0$$

In our problem $\nabla f = \begin{bmatrix} 4x \\ 4 \end{bmatrix}$ and $\nabla g = \begin{bmatrix} 1 \\ 1 \end{bmatrix}$. There fore from above $\begin{bmatrix} 4x \\ 4 \end{bmatrix} + \lambda \begin{bmatrix} 1 \\ 1 \end{bmatrix} = 0$. Solving gives $\lambda = -4$ and $x = 4/4 = 1$. Also we know $x + y + 3 = 0$.

$$\Rightarrow y = -x - 3 = -1 - 3 = -4.$$

Thus the optimal solution for computing the extremum of the function $f(x, y) = 2x^2 + 4y + 3$ is obtained as $(1, -4)$ and the corresponding value is $2–16+3 = -11$. But to test whether the obtained extremum is maximum or minimum is decided using the second derivative as shown below.

We have already shown that $\frac{d^2 f1(\alpha)}{d} \alpha^2 (at\ \alpha = \alpha_0) > 0$, where $f1(\alpha) = f(x(\alpha), y(\alpha))$ if the extremum point $\alpha = \alpha_0$ is minima point.

Using the functional dependencies as shown in the Fig. 5.2, $\frac{d^2 f1(\alpha)}{d}\alpha^2$ is computed as shown below.

$$\frac{df1}{d\alpha} = \frac{\partial f}{\partial x}\frac{\partial x}{\partial \alpha} + \frac{\partial f}{\partial y}\frac{\partial y}{\partial \alpha} = p(x(\alpha), y(\alpha))$$

$$\frac{d^2 f1(\alpha)}{d\alpha^2} = \frac{d\left(\frac{df1}{d\alpha}\right)}{d\alpha} = \frac{d(p(x(\alpha), y(\alpha))}{d\alpha} = \frac{\partial p}{\partial x}\frac{\partial x}{\partial \alpha} + \frac{\partial p}{\partial y}\frac{\partial y}{\partial \alpha}$$

$$= \frac{\partial x}{\partial \alpha}\frac{\partial\left(\frac{\partial f}{\partial x}\frac{\partial x}{\partial \alpha} + \frac{\partial f}{\partial y}\frac{\partial y}{\partial \alpha}\right)}{\partial x} + \frac{\partial y}{\partial \alpha}\frac{\partial\left(\frac{\partial f}{\partial x}\frac{\partial x}{\partial \alpha} + \frac{\partial f}{\partial y}\frac{\partial y}{\partial \alpha}\right)}{\partial y}$$

$$= \frac{\partial x}{\partial \alpha}\left(\frac{\partial^2 f}{\partial x^2}\right)\frac{\partial x}{\partial \alpha} + \frac{\partial x}{\partial \alpha}\frac{\partial f}{\partial x}\left(\frac{\partial^2 x}{\partial x \partial \alpha}\right) + \frac{\partial x}{\partial \alpha}\left(\frac{\partial^2 f}{\partial x \partial y}\right)\frac{\partial y}{\partial \alpha}$$
$$+ \frac{\partial x}{\partial \alpha}\frac{\partial f}{\partial y}\left(\frac{\partial^2 y}{\partial x \partial \alpha}\right) + \frac{\partial y}{\partial \alpha}\left(\frac{\partial^2 f}{\partial y^2}\right)\frac{\partial y}{\partial \alpha} + \frac{\partial y}{\partial \alpha}\frac{\partial f}{\partial y}\left(\frac{\partial^2 y}{\partial \alpha \partial y}\right)$$
$$+ \frac{\partial y}{\partial \alpha}\left(\frac{\partial^2 f}{\partial y \partial x}\right)\frac{\partial x}{\partial y} + \frac{\partial y}{\partial \alpha}\frac{\partial f}{\partial x}\left(\frac{\partial^2 x}{\partial y \partial \alpha}\right)$$
$$= \begin{bmatrix}\frac{\partial x}{\partial \alpha} & \frac{\partial y}{\partial \alpha}\end{bmatrix}\begin{bmatrix}\left(\frac{\partial^2 f}{\partial x^2}\right) & \frac{\partial^2 f}{\partial y \partial x} \\ \frac{\partial^2 f}{\partial y \partial x} & \frac{\partial^2 f}{\partial y^2}\end{bmatrix}\begin{bmatrix}\frac{\partial x}{\partial \alpha} \\ \frac{\partial y}{\partial \alpha}\end{bmatrix} + \begin{bmatrix}\frac{\partial f}{\partial x} & \frac{\partial f}{\partial y}\end{bmatrix}\begin{bmatrix}\frac{\partial^2 x}{\partial \alpha^2} \\ \frac{\partial^2 y}{\partial \alpha^2}\end{bmatrix}$$

Thus to satisfy the condition $\frac{d^2 f1(\alpha)}{d\alpha^2}(at\ \alpha = \alpha_0) > 0$

$$\begin{bmatrix}\frac{\partial x}{\partial \alpha} & \frac{\partial y}{\partial \alpha}\end{bmatrix}\begin{bmatrix}\left(\frac{\partial^2 f}{\partial x^2}\right) & \frac{\partial^2 f}{\partial y \partial x} \\ \frac{\partial^2 f}{\partial y \partial x} & \frac{\partial^2 f}{\partial y^2}\end{bmatrix}\begin{bmatrix}\frac{\partial x}{\partial \alpha} \\ \frac{\partial y}{\partial \alpha}\end{bmatrix} + \begin{bmatrix}\frac{\partial f}{\partial x} & \frac{\partial f}{\partial y}\end{bmatrix}\begin{bmatrix}\frac{\partial^2 x}{\partial \alpha^2} \\ \frac{\partial^2 y}{\partial \alpha^2}\end{bmatrix} (at\ \alpha = \alpha_0) > 0$$

Note that $\begin{bmatrix}\frac{\partial x}{\partial \alpha} & \frac{\partial y}{\partial \alpha}\end{bmatrix}$ $(at\ \alpha = \alpha_0)$ is the direction of tangent vector drawn at the point $\alpha = \alpha_0$ on the curve $g(x, y)$ and hence the point lies on the tangent vector drawn on the curve $g(x, y)$ at the point $\alpha = \alpha_0$.

Also $\lambda\ \frac{d^2 g(\alpha)}{d\alpha^2} = 0\ at(at\ \alpha = \alpha_0)$.

Expanding in the similar fashion as described above, it can be shown that

$$\lambda\begin{bmatrix}\frac{\partial x}{\partial \alpha} & \frac{\partial y}{\partial \alpha}\end{bmatrix}\begin{bmatrix}\left(\frac{\partial^2 g}{\partial x^2}\right) & \frac{\partial^2 g}{\partial y \partial x} \\ \frac{\partial^2 g}{\partial y \partial x} & \frac{\partial^2 g}{\partial y^2}\end{bmatrix}\begin{bmatrix}\frac{\partial x}{\partial \alpha} \\ \frac{\partial y}{\partial \alpha}\end{bmatrix} + \lambda\begin{bmatrix}\frac{\partial g}{\partial x} & \frac{\partial g}{\partial y}\end{bmatrix}\begin{bmatrix}\frac{\partial^2 x}{\partial \alpha^2} \\ \frac{\partial^2 y}{\partial \alpha^2}\end{bmatrix} (at\ \alpha = \alpha_0) = 0$$

Adding both the equation, we get

$$\begin{bmatrix} \frac{\partial x}{\partial \alpha} & \frac{\partial y}{\partial \alpha} \end{bmatrix} \begin{bmatrix} \left(\frac{\partial^2 f}{\partial x^2}\right) & \frac{\partial^2 f}{\partial y \partial x} \\ \frac{\partial^2 f}{\partial y \partial x} & \frac{\partial^2 f}{\partial y^2} \end{bmatrix} \begin{bmatrix} \frac{\partial x}{\partial \alpha} \\ \frac{\partial y}{\partial \alpha} \end{bmatrix} + \lambda \begin{bmatrix} \frac{\partial x}{\partial \alpha} & \frac{\partial y}{\partial \alpha} \end{bmatrix} \begin{bmatrix} \left(\frac{\partial^2 g}{\partial x^2}\right) & \frac{\partial^2 g}{\partial y \partial x} \\ \frac{\partial^2 g}{\partial y \partial x} & \frac{\partial^2 g}{\partial y^2} \end{bmatrix} \begin{bmatrix} \frac{\partial x}{\partial \alpha} \\ \frac{\partial y}{\partial \alpha} \end{bmatrix}$$

$$+ \begin{bmatrix} \frac{\partial f}{\partial x} & \frac{\partial f}{\partial y} \end{bmatrix} \begin{bmatrix} \frac{\partial^2 x}{\partial \alpha^2} \\ \frac{\partial^2 y}{\partial \alpha^2} \end{bmatrix} + \lambda \begin{bmatrix} \frac{\partial g}{\partial x} & \frac{\partial g}{\partial y} \end{bmatrix} \begin{bmatrix} \frac{\partial^2 x}{\partial \alpha^2} \\ \frac{\partial^2 y}{\partial \alpha^2} \end{bmatrix} \quad (at\ \alpha = \alpha_0) > 0$$

We have already shown that

$$\begin{bmatrix} \frac{\partial f}{\partial x} & \frac{\partial f}{\partial y} \end{bmatrix} + \lambda \begin{bmatrix} \frac{\partial g}{\partial x} & \frac{\partial g}{\partial y} \end{bmatrix} = 0$$

and hence we get the following condition.

$$\begin{bmatrix} \frac{\partial x}{\partial \alpha} & \frac{\partial y}{\partial \alpha} \end{bmatrix} \begin{bmatrix} \left(\frac{\partial^2 f}{\partial x^2}\right) + \lambda \left(\frac{\partial^2 g}{\partial x^2}\right) & \frac{\partial^2 f}{\partial y \partial x} + \lambda \frac{\partial^2 g}{\partial y \partial x} \\ \frac{\partial^2 f}{\partial y \partial x} + \lambda \frac{\partial^2 g}{\partial y \partial x} & \frac{\partial^2 f}{\partial y^2} + \lambda \frac{\partial^2 g}{\partial y^2} \end{bmatrix} \begin{bmatrix} \frac{\partial x}{\partial \alpha} \\ \frac{\partial y}{\partial \alpha} \end{bmatrix} \quad (at\ \alpha = \alpha_0) > 0$$

In practice representing the function $g(x, y)$ in parametric form is not easy like the one used in the example. But the point $\begin{bmatrix} \frac{\partial x}{\partial \alpha} & \frac{\partial y}{\partial \alpha} \end{bmatrix}$ always lies on the tangent of the curve $g(x, y)$ drawn at the point (x_0, y_0) and hence to confirm whether the obtained point is minima, we have to test whether the modified Hessian matrix $\begin{bmatrix} \left(\frac{\partial^2 f}{\partial x^2}\right) + \lambda \left(\frac{\partial^2 g}{\partial x^2}\right) & \frac{\partial^2 f}{\partial y \partial x} + \lambda \frac{\partial^2 g}{\partial y \partial x} \\ \frac{\partial^2 f}{\partial y \partial x} + \lambda \frac{\partial^2 g}{\partial y \partial x} & \frac{\partial^2 f}{\partial y^2} + \lambda \frac{\partial^2 g}{\partial y^2} \end{bmatrix}$ is positive definite restricted to the points on the tangent of the curve $g(x, y)$ drawn at the point (x_0, y_0).

The modified Hessian matrix to be tested is given as

$$\begin{bmatrix} \frac{d^2 f}{dx^2} + \lambda \frac{d^2 g}{dx^2} & \frac{d^2 f}{dx\,dy} + \lambda \frac{d^2 g}{dx\,dy} \\ \frac{d^2 f}{dydx} + \lambda \frac{d^2 g}{dx\,dy} & \frac{d^2 f}{dy^2} + \lambda \frac{d^2 g}{dx^2} \end{bmatrix} \quad \text{with } \lambda = -4, \text{ as found earlier}$$

$$= \begin{bmatrix} 4 + (-4) * 0 & 0 + 0 \\ 0 + (-4) * 0 & 0 + 0 \end{bmatrix} = \begin{bmatrix} 4 & 0 \\ 0 & 0 \end{bmatrix} = A\ (say)$$

The obtained optimal solution $x = 1, y = -4$ (described earlier) corresponds to the minimal point if the modified Hessian matrix (A) as shown above is the positive definite restricted to the points on the tangent drawn on the curve $g(x, y)$ at the point $(1, -4)$.

The equation of the tangent drawn on the curve $g(x, y)$ at the point $(1, -4)$ is obtained as the set of points $(z1, z2)$ satisfying the condition $\nabla g^T [z1-1 \;\; z2+4] = 0$. Note that ∇g^T is computed at $(1, -4)$.

For the function $g(x, y) = x+y+3 = 0$, the gradient vector at the point $(1, -4)$ is found as follows.

$$\begin{bmatrix} \frac{\partial g}{\partial x} \\ \frac{\partial g}{\partial y} \end{bmatrix}^T = \begin{bmatrix} 1 \\ 1 \end{bmatrix}$$

Therefore the equation of the tangent drawn at the point $(1, -4)$ on the curve $g(x, y) = x + y + 3 = 0$ is obtained as $\begin{bmatrix} 1 \\ 1 \end{bmatrix}^T [z1 - 1 \; z2 + 4] = z1 + z2 + 3 = 0$.

The set of points $(v, -3-v) \forall \, v \in \mathbb{R}$ is the parametric representation of the above equation that lies on the tangent drawn on the curve $g(x, y)$ at $(1, -4)$.

Also, the matrix A is said to be positive definite restricted to the points that lies on the tangent drawn on the curve $g(x, y)$ at the point $(1, -4)$. if $u^T A u > 0, \;\; \forall u \in$ points as described above.

$$u^T Au = [v \quad -3-v] \begin{bmatrix} 4 & 0 \\ 0 & 0 \end{bmatrix} \begin{bmatrix} v \\ -3-v \end{bmatrix} = [4v \quad 0] \begin{bmatrix} v \\ -3-v \end{bmatrix} = 4v^2$$

which is greater than or equal to zero and hence the point obtained is the saddle point.

5.2 Extension to Constrained Optimization Technique to Higher Dimensional Space with Multiple Constraints

In general the problem discussed above can be extended to more than two variables and more than one constraints as shown below.

Minimize $f([x1 \; x2 \; x3 \ldots xn])$, subject to the constraints

$$f([x1\, x2\, x3 \ldots xn]) \quad g2([x1\, x2\, x3 \ldots xn]) = 0 \ldots gm([x1\, x2\, x3 \ldots xn]) = 0.$$

Create the Lagrangean function

$$\begin{aligned} &L(f, g1, g2, g3) \\ &= f([x1\, x2\, x3 \ldots xn]) + \lambda_1 g_1([x1\, x2\, x3 \ldots xn]) \\ &\quad + \lambda_2 g_2([x1\, x2\, x3 \ldots xn]) + \ldots \lambda_m g_m([x1\, x2\, x3 \ldots xn]) \end{aligned}$$

Differentiating the above equation with respect to $x1 \;\; x2 \;\; x3 \ldots xn, \lambda_1, \lambda_2, \ldots \lambda_m$ and equate to zero gives the extremum point $(x_{01} \; x_{02} \; x_{03} \ldots x_{0n})$ (say).

The above set of equations can also be obtained using the following equation

$$\nabla f + [\nabla g_1 \nabla g_2 \ldots \nabla g_{1m}] \begin{bmatrix} \lambda_1 \\ \lambda_2 \\ \ldots \\ \lambda_m \end{bmatrix} = 0.$$

Modified Hessian matrix is used to test whether the obtained extremum point thus obtained is minimum or not as shown below.

The Modified Hessian matrix is as shown in below

$$\begin{bmatrix} \frac{d^2 f}{dx_1{}^2} + \lambda_1 \frac{d^2 g_1}{dx_1{}^2} + \lambda_2 \frac{d^2 g_2}{dx_1{}^2} + \lambda_3 \frac{d^2 g_3}{dx_1{}^2} & \frac{d^2 f}{dx_1 x_2} + \lambda_1 \frac{d^2 g_1}{dx_1 x_2} + \lambda_2 \frac{d^2 g_2}{dx_1 x_2} + \lambda_3 \frac{d^2 g_3}{dx_1 x_2} & \ldots & \frac{d^2 f}{dx_1 x_n} + \lambda_1 \frac{d^2 g_1}{dx_1 x_n} + \lambda_2 \frac{d^2 g_2}{dx_1 x_n} + \lambda_3 \frac{d^2 g_3}{dx_1 x_n} \\ \frac{d^2 f}{dx_2 x_1} + \lambda_1 \frac{d^2 g_1}{dx_2 x_1} + \lambda_2 \frac{d^2 g_2}{dx_1 x_1} + \lambda_3 \frac{d^2 g_3}{dx_1 x_1} & \frac{d^2 f}{dx_2{}^2} + \lambda_1 \frac{d^2 g_1}{dx_2{}^2} + \lambda_2 \frac{d^2 g_2}{dx_2{}^2} + \lambda_3 \frac{d^2 g_3}{dx_2{}^2} & \ldots & \frac{d^2 f}{dx_2 x_n} + \lambda_1 \frac{d^2 g_1}{dx_2 x_n} + \lambda_2 \frac{d^2 g_2}{dx_3 x_n} + \lambda_3 \frac{d^2 g_3}{dx_n x_n} \\ \ldots & \ldots & \ldots & \ldots \\ \frac{d^2 f}{dx_n x_1} + \lambda_1 \frac{d^2 g_1}{dx_n x_1} + \lambda_2 \frac{d^2 g_2}{dx_n x_1} + \lambda_3 \frac{d^2 g_3}{dx_n x_1} & \frac{d^2 f}{dx_n x_2} + \lambda_1 \frac{d^2 g_1}{dx_n x_2} + \lambda_2 \frac{d^2 g_2}{dx_n x_2} + \lambda_3 \frac{d^2 g_3}{dx_n x_2} & \ldots & \frac{d^2 f}{x_n{}^2} + \lambda_1 \frac{d^2 g_1}{dx_n{}^2} + \lambda_2 \frac{d^2 g_2}{dx_n{}^2} + \lambda_3 \frac{d^2 g_3}{dx_n{}^2} \end{bmatrix}$$

If the Modified Hessian matrix as mentioned above is positive definite restricted to the points on the tangent plane P as defined below.

Tangent plane 'P' drawn at the optimal point $(x_{01}\ x_{02}\ x_{03} \ldots x_{0n})$ on the surface defined by the equations $g_1(x1, x2, ..xn)$, $g_2(x1, x2, ..xn)$ and $g_3(x1, x2, ..xn)$ is defined as the set of points 'y' satisfying the equation.

$$[\nabla g_1\ \nabla g_2\ \nabla g_3]^T\ (computed\ at\ [x_{01}\quad x_{02}\ x_{03}\quad \ldots x_{0n}]) \left(\underline{y} - \begin{bmatrix} x_{01} \\ x_{02} \\ x_{03} \\ x_{04} \\ \ldots \\ x_{0n} \end{bmatrix} \right) = 0$$

Example 5.1. Minimize the function $f(x1, x2, x3) = 2x1^2 + 4 * x2 + 2 * x3 + 2$. Subject to the constraint $g1(x1, x2, x3) = x1 + x2 - 4 = 0,\ g2(x1, x2, x3) = x1 + x3 - 5 = 0$

$$\nabla f + [\nabla g_1 \nabla g_2 \ldots \nabla g_{1m}] \begin{bmatrix} \lambda_1 \\ \lambda_2 \\ \ldots \\ \lambda_m \end{bmatrix} = 0$$

$$\Rightarrow \begin{bmatrix} 4x1 \\ 4 \\ 2 \end{bmatrix} + \begin{bmatrix} 1 & 1 \\ 1 & 0 \\ 0 & 1 \end{bmatrix} \begin{bmatrix} \lambda_1 \\ \lambda_2 \end{bmatrix} = 0$$

$$\Rightarrow \begin{bmatrix} 4x1 + \lambda_1 + \lambda_2 = 0 \\ 4 + \lambda_1 + \lambda_2 = 0 \\ 2 + \lambda_2 = 0 \end{bmatrix}$$

Solving the above equation gives $\lambda_2 = -2, \lambda_1 = -2, x1 = 1, x2 = 3, x3 = 4$

The optimum value is (1,3,4) and the corresponding function value is 24.

To test whether the obtained solution is minima or not is done using modified Hessian matrix as given below.

$$\begin{bmatrix} 4 & 0 & 0 \\ 0 & 0 & 0 \\ 0 & 0 & 0 \end{bmatrix}$$

The optimal point thus obtained is minima if the modified Hessian matrix obtained is positive definite restricted to the points on the tangent plane of the surface defined by the equation $g1(x1, x2, x3) = x1 + x2 - 4 = 0,\ g2(x1, x2, x3) = x1 + x3 - 5 = 0$

The equation of the tangent as described above is obtained as follows.

$$[\nabla g1 \quad \nabla g2]^T \begin{bmatrix} y1 - 1 \\ y2 - 3 \\ y3 - 4 \end{bmatrix} = 0$$

$$\Rightarrow \begin{bmatrix} 1 & 1 & 0 \\ 1 & 0 & 1 \end{bmatrix} \begin{bmatrix} y1 - 1 \\ y2 - 3 \\ y3 - 4 \end{bmatrix} = 0$$

$$\Rightarrow \begin{bmatrix} 1 & 1 & 0 \\ 1 & 0 & 1 \end{bmatrix} \begin{bmatrix} y1 \\ y2 \\ y3 \end{bmatrix} = \begin{bmatrix} 4 \\ 5 \end{bmatrix}$$

$$y1 + y2 = 4$$
$$y1 + y3 = 5$$

Let $y1 = \alpha \quad y2 = 4 - \alpha$ and $y3 = 5 - \alpha$

Thus the tangent plane is defined as the vector space that is spanned by the column vectors as described below.

$$\begin{bmatrix} \alpha \\ 4-\alpha \\ 5-\alpha \end{bmatrix}, \forall \alpha \in \mathbb{R}$$

To check whether the matrix $\begin{bmatrix} 4 & 0 & 0 \\ 0 & 0 & 0 \\ 0 & 0 & 0 \end{bmatrix}$ is negative definite restricted to the vector of the form $\begin{bmatrix} \alpha \\ 4-\alpha \\ 5-\alpha \end{bmatrix}$ is tested as described below.

$$[\alpha \quad 4-\alpha \quad 5-\alpha] \begin{bmatrix} 4 & 0 & 0 \\ 0 & 0 & 0 \\ 0 & 0 & 0 \end{bmatrix} \begin{bmatrix} \alpha \\ 4-\alpha \\ 5-\alpha \end{bmatrix} = 4\alpha^2 \geq 0 \forall \alpha \in \mathbb{R}$$

Hence the obtained extremal point is the saddle point.

5.3 Positive Definite Test of the Modified Hessian Matrix Using Eigen Value Computation

If all the Eigen values of the matrix modified Hessian matrix computed at the extremum point are positive, the matrix is said to be positive definite matrix and the corresponding extremum point is minima. If all the Eigen values of the matrix modified Hessian matrix computed at the extremum point are negative, the matrix is said to be Negative definite and the corresponding extremum point belongs to maxima.

Proof. As Hessian matrix $A = \begin{bmatrix} a1 & a2 & a3 \\ a2 & a4 & a5 \\ a3 & a5 & a6 \end{bmatrix}$ (say) is the symmetric matrix, it is diagonalizable (i.e.) the matrix can be represented as $A = EDE^H$. Also the Eigen values of the matrix A are real numbers and the Eigen vectors are orthonormal to each other. (Refer Chapter 4)

$$[x1\ \ x2\ \ x3]\begin{bmatrix} a1 & a2 & a3 \\ a2 & a4 & a5 \\ a3 & a5 & a6 \end{bmatrix}\begin{bmatrix} x1 \\ x2 \\ x3 \end{bmatrix}$$

$$= [x1\ x2\ x3]\begin{bmatrix} e11 & e21 & e31 \\ e12 & e22 & e32 \\ e13 & e23 & e33 \end{bmatrix}\begin{bmatrix} \lambda_1 & 0 & 0 \\ 0 & \lambda_2 & 0 \\ 0 & 0 & \lambda_3 \end{bmatrix}\begin{bmatrix} e11 & e12 & e13 \\ e21 & e22 & e23 \\ e31 & e32 & e33 \end{bmatrix}\begin{bmatrix} x1 \\ x2 \\ x3 \end{bmatrix}$$

$$= \lambda_1(x1 * e11 + x2 * e12 + x3 * e13)^2$$
$$+\lambda_2(x1 * e21 + x2 * e22 + x3 * e23)^2$$
$$+\lambda_3(x1 * e31 + x2 * e32 + x3 * e33)^2$$

Thus to test whether the matrix A is positive definite, the requirement is

$$\lambda_1(x1 * e11 + x2 * e12 + x3 * e13)^2 + \lambda_2(x1 * e21 + x2 * e22 + x3 * e23)^2$$
$$+\lambda_3(x1 * e31 + x2 * e32 + x3 * e33)^2 > 0$$

This implies all the Eigen values should be greater than zero.

Example 5.2. Consider the problem of minimizing the function $f(x, y) = 2x^2 + 4y + 3$ subject to the constraint $g(x, y) = x + y + 3 = 0$.

We have already shown that the extremal point is $(1, -4)$ with respect to the xy co-ordinate system. Now shifted co-ordinate system PQ co-ordinate system is framed such that the extremal point is (0,0) with respect to the new co-ordinate system.

$$\Rightarrow \mathrm{P} = \mathrm{X} - 1, \mathrm{Q} = \mathrm{Y} + 4$$

Thus the function $g(x, y) = x + y + 3 = 0$ is rewritten with respect to the new co-ordinate system as shown below.

$$u(p, q) = p + 1 + q - 4 + 3 = 0$$
$$\Rightarrow u(\mathrm{p}, \mathrm{q}) = \mathrm{p} + \mathrm{q} = 0$$

The parametric representation of the above equation is represented as the set of points of the form $(\beta, -\beta)$. The equation of the tangent plane with respect to the new co-ordinate system is the set of points $y = [y1\ y2]^T$ such that $\nabla u^T\ y = 0$

$$\Rightarrow [1 \quad 1]\begin{bmatrix} y1 \\ y2 \end{bmatrix} = 0.$$

In other words the set of points describing the tangent plane is in the Null space of the matrix $[1 \quad 1]$, which is the represented as $\begin{bmatrix} \alpha \\ -\alpha \end{bmatrix}$, *where* $\alpha \in \mathbb{R}$.

The Modified Hessian matrix is found at the point (0,0) with the new co-ordinate system and is displayed below. The equation $f(x, y) = 2x^2+4y+3$ in the modified co-ordinate system is given as

$$v(p, q) = 2 * (p + 1)^2 + 4 * (q + 5) + 3or$$
$$= 2(p^2 + 1 + 2p) + 4q + 20 + 3a = 2p^2 + 4p + 4p + 25$$

Modified Hessian matrix is $\begin{bmatrix} 4 & 0 \\ 0 & 0 \end{bmatrix}$ (Note that the value of $\lambda = -4$ which is same as the one calculated with the previous co-ordinate system). To test whether the above Hessian matrix is positive definite or not restricted to the tangent plane as described above is as shown below.

$[\alpha \quad -\alpha] \begin{bmatrix} 4 & 0 \\ 0 & 0 \end{bmatrix} \begin{bmatrix} \alpha \\ -\alpha \end{bmatrix} = 4\alpha^2 \geq 0$ and hence the extremal point obtained is the indeterminate point.

Trick to test whether the Hessian matrix 'H' is positive definite restricted to the points on the tangent plane as described above.

Any vector in the tangent plane can be represented as the linear combinations of the basis of the Null space of the matrix ∇u^T (see above). The Eigen basis vector E1,E2,E3 (say) are arranged in column form to obtain Eigen matrix E. Any vector in the tangent plane space can be represented as the linear combinations of the Eigen vectors as described below. Thus $[E1 \;\; E2] \begin{bmatrix} p1 \\ p2 \end{bmatrix}$ is the arbitrary point in the tangent plane as described above.

Thus the Hessian matrix H is positive definite restricted to the points on the tangent plane (as described above) if $[p1 \;\; p2]E^T \; HE \begin{bmatrix} p1 \\ p2 \end{bmatrix} > 0 \forall p1, p2 \in \mathbb{R}$.

Thus if the Eigen values of the matrix $E^T \; HE$ is greater than 0, the matrix is positive definite matrix.

In the above example, Eigen values of the matrix $\left[\begin{bmatrix} \frac{1}{\sqrt{2}} & -\frac{1}{\sqrt{2}} \end{bmatrix} \begin{bmatrix} 4 & 0 \\ 0 & 0 \end{bmatrix} \begin{bmatrix} \frac{1}{\sqrt{2}} \\ -\frac{1}{\sqrt{2}} \end{bmatrix}\right] =$ [2] shave to be found. The Eigen value is positive and hence the Extremum point obtained is minimum.

Example 5.3. Consider the problem described in Example 5.1.

Minimize the function $f(x1, x2, x3) = 2x1^2+4*x2+2*x3+2$. Subject to the constraint $g1(x1, x2, x3) = x1 + x2 - 4 = 0$, $g2(x1, x2, x3) = x1 + x3 - 5 = 0$.

The optimal point is obtained as (1,3,4) with respect to the co-ordinate system $(x1, x2, x3)$. The modified co-ordinate system $(z1, z2, z3)$ is obtained as $z1=x1-1$; $z2 = x2 - 3$; $z3 = x3 - 4$;

The Modified equations corresponding to the functions $g1, g2,$ *and* f are $u1, u2$ *and* v respectively which are displayed below

$$\begin{aligned} u1(z1, z2, z3) &= z1 + 1 + z2 + 3 - 4 = 0 = z1 + z2 = 0 \\ u2(z1, z2, z3) &= z1 + 1 + z3 + 4 - 5 = 0 = z1 + z3 = 0 \\ v(z1, z2, z3) &= 2(z1 + 1)^2 + 4 * (z2 + 3) + 2 * (z3 + 4) + 2 \\ &= 2(z1^2 + 1 + 2z1) + 4z2 + 12 + 2z3 + 8 + 2 \\ &= 2z1^2 + 4z1 + 4z2 + 2z3 + 24 \end{aligned}$$

The Modified Hesian matrix 'H' at the point (0,0,0) with respect to the modified co-ordinate system is given below. $\begin{bmatrix} 4 & 0 & 0 \\ 0 & 0 & 0 \\ 0 & 0 & 0 \end{bmatrix}$. (Note that the value of $\lambda 1 = -2$ and $\lambda 2 = -2$, which are same as that of the one calculated with the previous co-ordinate system).

The Equation of the tangent plane passing through the point (0,0,0) in the new co-ordinate system is described as the set of points (p,q,r) satisfies the following condition.

$$[\nabla g1 \ \nabla g2]^T \begin{bmatrix} p \\ q \\ r \end{bmatrix} = 0$$

$$\Rightarrow \begin{bmatrix} 1 & 1 & 0 \\ 1 & 0 & 1 \end{bmatrix} \begin{bmatrix} p \\ q \\ r \end{bmatrix} = 0$$

The set of points on the tangent plane is given as the null space of the matrix $\begin{bmatrix} 1 & 1 & 0 \\ 1 & 0 & 1 \end{bmatrix}$. The Eigen basis of the null space is represented as B = $\begin{bmatrix} \frac{1}{\sqrt{3}} \\ -\frac{1}{\sqrt{3}} \\ -\frac{1}{\sqrt{3}} \end{bmatrix}$

The Eigen value of the matrix $E^T H E = [1 \quad -1 \quad -1] \begin{bmatrix} 4 & 0 & 0 \\ 0 & 0 & 0 \\ 0 & 0 & 0 \end{bmatrix} \begin{bmatrix} 1 \\ -1 \\ -1 \end{bmatrix} = [4]$ is 4 which is greater than zero and hence the obtained minima point is the minima point.

Example 5.4. Maximize the function $x_1x_2 + x_2x_3 + x_1x_3$.subject to the constraint $x_1 + x_2 + x_3 = 3$.

The Lagrangean function is obtained as $x_1x_2 + x_2x_3 + x_1x_3 + \lambda(x_1 + x_2 + x_3 - 3) = 0$.

The following set of equations are obtained by computing by differentiating the Lagrangean function with respect to $(x1, x2, x3, \lambda)$ and equate to zero.

$$x_2 + x_3 + \lambda = 0;\ x_1 + x_3 + \lambda = 0; x_1 + x_2 + \lambda = 0; x_1 + x_2 + x_3 = 3$$

Solving the above equation gives $\lambda = -2, x_1 = x_2 = x_3 = 1$

The modified co-ordinate system (z_1, z_1, z_1) are obtained as $z_1 = x_1 - 1; z_2 = x_2 - 1; z_3 = x_3 - 1$.

The modified equation corresponding to f and g is as shown below.

$$\begin{aligned} v(z_1, z_2, z_3) &= (z_1 + 1)(z_2 + 1) + (z_2 + 1)(z_3 + 1) + (z_1 + 1)(z_3 + 1) \\ &= z_1 z_2 + z_2 z_3 + z_3 z_1 + z_1 + z_2 + z_3 + 3 \\ u(z_1, z_2, z_3) &= z_1 + 1 + z_2 + 1 + z_3 + 1 = z_1 + z_2 + z_3 + 3 = 0 \end{aligned}$$

The Modified Hessian matrix at (0,0,0) in the new co-ordinate system is computed as $\begin{bmatrix} 0 & 1 & 1 \\ 1 & 0 & 1 \\ 1 & 1 & 0 \end{bmatrix}$

The Equation of the tangent plane with the modified co-ordinate system is set of points (p, q, r) satisfying the condition $\nabla u^T \begin{bmatrix} p \\ q \\ r \end{bmatrix} = 0$

This implies the set of points form the null space of the matrix [1 1 1]

The basis of the null space of the matrix [1 1 1] is given as $\begin{bmatrix} -0.5774 & -0.5774 \\ 0.7887 & -0.2113 \\ -0.2113 & 07887 \end{bmatrix}$.

The Eigen values of the matrix $\begin{bmatrix} -0.5774 & -0.5774 \\ 0.7887 & -0.2113 \\ -0.2113 & 07887 \end{bmatrix}^T \begin{bmatrix} 0 & 1 & 1 \\ 1 & 0 & 1 \\ 1 & 1 & 0 \end{bmatrix}$ $\begin{bmatrix} -0.5774 & -0.5774 \\ 0.7887 & -0.2113 \\ -0.2113 & 07887 \end{bmatrix} = \begin{bmatrix} -1 & 0 \\ 0 & -1 \end{bmatrix}$ are given as -1 and -1 which are less than zero and hence the obtained extremum point is the maxima point.

5.4 Constrained Optimization with Complex Numbers

Consider the problem of minimizing the function $f(x1, x2, x3)$ *where* $x1, x2, x3 \in \mathbb{C}$ subject to the constraint

$$g(x1, x2, x3) = 0, \textit{where } x1, x2, x3 \in \mathbb{C}$$

The problem can be viewed as maximizing both real and imaginary part of the function. Hence the Lagrangean function is given as

$$f(x1, x2, x3) + \lambda g(x1, x2, x3)$$

The equation can be written as below treating the complex Lagrange multiplier defined as $\lambda = \lambda_1 + j\lambda_2$.

$$f(x1, x2, x3) + Re\left(\bar{\lambda} g(x1, x2, x3)\right) = 0$$

Differentiating the above equation with respect to $\overline{x1}, \overline{x2}, \overline{x3}$ *and* $\bar{\lambda}$ and equate to zero gives the extremal point for the above problem.

Note: Differentiation of the function with respect the complex number is defined as shown below.

$$\frac{\partial}{\partial \overline{x1}} = \frac{1}{2}\left(\frac{\partial}{\partial x_{11}} + j\frac{\partial}{\partial x_{12}}\right) \text{ and } \frac{\partial}{\partial x1} = \frac{1}{2}\left(\frac{\partial}{\partial x_{11}} - j\frac{\partial}{\partial x_{12}}\right)$$

Properties of the complex differentiation

1. $\frac{\partial x1}{\partial \overline{x1}} = 0$
2. $\frac{\partial x1}{\partial x1} = 1$
3. $\frac{\partial (A^H x1)}{\partial \overline{x1}} = 0$
4. $\frac{\partial (A^H x1)}{\partial x1} = \bar{A}$
5. $\frac{\partial (z^H A)}{\partial \bar{z}} = A$
6. $\frac{\partial Re(z^H A)}{\partial \bar{z}} = \frac{1}{2}A$

5.5 Dual Optimization Problem

Consider the problem of minimizing the function $f(x1, x2, x3)$, subject to the constraint $g1(x1, x2, x3) = 0$. Framing the Lagrange equation we get, $L(X, \lambda) = f(x1, x2, x3) + \lambda g1(x1, x2, x3)$. Differentiate the above equation with respect to $x1, x2, x3, \lambda$ and equate to zero. Solve for $x1, x2, x3$ in terms λ and substitute in the equation $L(X, \lambda)$ to obtain the function $h(\lambda)$. Thus function thus obtained is called dual optimization problem for the above mentioned constrained optimization problem. Thus the Dual problem is as shown below.

Maximize the function $h(\lambda)$ without constraints. (i.e.) Unconstrained optimization problem.

Example 5.5. Consider the problem of minimizing the function $f(x1, x2, x3) = x1x2 + x2x3 + x3x1 + x1x2x3 - 4 = 0$. Subject to the following constraints $g1(x1, x2, x3) = x1 + x2 + x3 - 3 = 0$.

Framing the Lagrange equation we get

$$L(X,\lambda) = x1x2 + x2x3 + x3x1 + x1x2x3 - 4 + \lambda(x1 + x2 + x3 - 3)$$

Differentiating the above equation with respect to $x1, x2, x3$ and equate to zero and solving for $x1, x2, x3$ in terms of λ, we get

$$x1 = \frac{1}{2} - \frac{1}{2}(7 + 2\lambda)^{\frac{1}{2}},\ x2 = \frac{1}{2} - \frac{1}{2}(7 + 2\lambda)^{\frac{1}{2}},\ x3 = 2 + (7 + 2\lambda)^{1/2}$$

Substituting the above equation in the Lagrange equation $L(X,\lambda)$, we get the Dual problem Maximizing the $\left(\frac{1}{2} - \frac{1}{2}(7 + 2\lambda)^{\frac{1}{2}}\right)\left(\frac{1}{2} - \frac{1}{2}(7 + 2\lambda)^{\frac{1}{2}}\right)$ without any constraints. (i.e.)Unconstrained optimization.

5.6 Kuhn-Tucker Conditions

Consider the optimization problem as shown below.

Minimize the function $f(x1, x2, x3)$. Subject to the constraints that

$$g1(x1,x2,x3) = 0\ g2(x1,x2,x3) = 0\ h1(x1,x2,x2) \leq 0$$
$$h2(x1,x2,x3) \leq 0$$

The Lagrangean function is framed with the Lagrangean multipliers $\lambda_1, \lambda_2, \mu_1, \mu_2$ as shown below.

$$f(x1,x2,x3) + \lambda_1 g1(x1,x2,x3) + \lambda_2 g2(x1,x2,x3)$$
$$+\mu_1 h1(x1,x2,x3) + \mu_2 h2(x1,x2,x3) = 0$$

Differentiating the above equation with respect to $x1, x2, x3$ and equate to zero to obtain the three equations. Also, The Lagrange multiplier used for inequality constraints satisfies the following conditions

$$\mu_1 h1(x1,x2,x3) + \mu_2 h2(x1,x2,x3) = 0$$
$$\mu_1 \geq 0, \mu_2 \geq 0$$

Along with this, the two equations $g1(x1,x2,x3) = 0\ g2(x1,x2,x3) = 0$ are used to obtain the extremal point.1

Example 5.6. Minimize the function

$f(x1,x2,x3) = x1x2 + x2x3 + x3x1 + x1x2x3 - 4 = 0$. Subject to the following constraints $g1(x1,x2,x3) = x1 + x2 + x3 - 3 = 0$

$$h1(x1,x2,x3) = x1 - x2 - x3 \leq 0, h2(x1,x2,x3) = x1 + x2 - 2x3 \leq 0$$

Construct the Lagrangean function

$$x1x2 + x2x3 + x3x1 + x1x2x3 + \lambda_1(x1 + x2 + x3 - 4) \\ +\mu_1(x1 - x2 - x3) + \mu_2(x1 + x2 + 2x3) = 0$$

Differentiating with respect to $x1, x2, x3$ and equate to zero, the following equations are obtained

$$x2 + x3 + x2x3 + \lambda_1 + \mu_1 + \mu_2 = 0$$
$$x1 + x3 + x1x3 + \lambda_1 - \mu_1 + \mu_2 = 0$$
$$x2 + x1 + x1x2 + \lambda_1 - \mu_1 - 2\mu_2 = 0$$

Also $\mu_1(x1 - x2 - x3) + \mu_2(x1 + x2 - 2x3) = 0$

Along with above equation $g1(x1, x2, x3) = x1 + x2 + x3 - 3 = 0$ is used to obtain the extremal points as given below.

Case 1: $\mu_1 = 0$. The following solutions are obtained

$$(x1, x2, x3, \lambda_1, \mu_1, \mu_2) \\ = (-1, 0, 4, 1, 0, 0), (3, 0, 0, -3, 0, 0), (1, 1, 1, -3, 0, 0), (-1, 5, -1, 1, 0, 0)$$

All the solutions are valid as $\mu_1 \geq 0, \textit{and } \mu_2 \geq 0$

Case 2: $\mu_2 = 0$. The following solutions are obtained

$$(x1, x2, x3, \lambda_1, \mu_1, \mu_2) = (-1, 0, 4, 1, 0, 0)(3, 0, 0, -3, 0, 0) \\ \left(\frac{3}{2}, \frac{3}{4}, \frac{3}{4}, -\frac{99}{32}, \frac{9}{32}, 0\right)(1, 1, 1, -3, 0, 0)(-1, 5, -1, 1, 0, 0)$$

All the solutions obtained above are valid as $\mu_1 \geq 0, \textit{and } \mu_2 \geq 0$.

Chapter 6
Matlab Illustrations

6.1 Generation of Multivariate Gaussian Distributed Sample Outcomes with the Required Mean Vector 'M_Y' and Covariance Matrix 'C_Y'

Consider the transfer of random variables X = R cos Θ and Y = R sin Θ, where 'Θ' is uniformly distributed between 0 and 2π and 'R' is Rayleigh density function as described below. Also R and 'Θ' are independent random variables.

The Rayleigh density function is given as follows

$$f_R(r) = \frac{r}{\sigma^2} e^{-\frac{r^2}{2\sigma^2}} \forall r \geq 0$$
$$= 0, elsewhere$$

The corresponding distribution function is given as follows.

$$F_R(r) = 1 - e^{-\frac{r^2}{2\sigma^2}} \forall r \geq 0$$
$$= 0, elsewhere$$

The joint density function of X and Y represented as $f_{XY}(x,\ y)$ is obtained using Jacobian as follows.

$$f_{XY}(x, y) = \frac{1}{|J|} f_{R\Theta}(r(x, y), \Theta(x, y))$$

The solution for the above set of equation gives $R = \sqrt{X^2 + Y^2}$ and $\Theta = tan^{-1}\left(\frac{Y}{X}\right)$. The Jacobian matrix at the solution is obtained as follows.

$$\begin{bmatrix} \frac{\partial X}{\partial R} & \frac{\partial X}{\partial \Theta} \\ \frac{\partial Y}{\partial R} & \frac{\partial Y}{\partial \Theta} \end{bmatrix} at \left(R = \sqrt{X^2 + Y^2}, \Theta = tan^{-1}\left(\frac{Y}{X}\right)\right) \text{ is obtained as}$$
$$\text{R} = \sqrt{X^2 + Y^2}$$

E.S. Gopi, *Mathematical Summary for Digital Signal Processing Applications with Matlab*, DOI 10.1007/978-90-481-3747-3_6,

Thus the joint density function of

$$\begin{aligned} f_{XY}(x,y) &= \frac{1}{\sqrt{X^2+Y^2}} f_{R\Theta}\left(\sqrt{x^2+y^2}, tan^{-1}\left(\frac{y}{x}\right)\right) \\ &= \frac{1}{\sqrt{X^2+Y^2}} f_R\left(\sqrt{x^2+y^2}\right) f_\Theta\left(tan^{-1}\left(\frac{y}{x}\right)\right) \\ &= \frac{1}{\sqrt{X^2+Y^2}} \frac{\left(\sqrt{x^2+y^2}\right)}{\sigma^2} e^{-\frac{(x^2+y^2)}{2\sigma^2}} \frac{1}{2\pi} \\ &= \frac{1}{\sqrt{2\pi\sigma^2}} e^{-\frac{x^2}{2\sigma^2}} \frac{1}{\sqrt{2\pi\sigma^2}} e^{-\frac{y^2}{2\sigma^2}} = f_X(x) f_Y(y) \end{aligned}$$

Note that the density functions $f_X(x)$, $f_Y(y)$ obtained above are the independent Gaussian density function with variance σ^2 and mean zero.

Thus the steps involved in generating the sample outcomes that is Gaussian distributed with mean = 0 and variance = σ^2 is summarized below.

Step 1: Generation of sample outcomes that is Rayleigh distributed from the uniformly distributed sample outcomes

Consider the Uniformly distributed random variable be 'U' and Rayleigh distributed random variable be 'R'. Also Let R = g(U)

$$F_R(r) = P(R \leq r) = P(\mathrm{g(U)} \leq \mathrm{r}) = \mathrm{P(U \leq g^{-1}(r))} = F_U(\mathrm{g^1(r)}) = \mathrm{g^{-1}(r)}$$

(As U is the uniformly distributed function).

$$\begin{aligned} &\Rightarrow F_R(\mathrm{r}) = \mathrm{g^{-1}(r)} \\ &\Rightarrow 1 - e^{-\frac{r^2}{2\sigma^2}} = \mathrm{g^{-1}(r)} = \mathrm{u} \end{aligned}$$

Solving for 'r', we get $e^{-\frac{r^2}{2\sigma^2}} = 1 - u \Rightarrow \frac{r^2}{2\sigma^2} = \ln\left(\frac{1}{1-u}\right)$

$$\Rightarrow \mathrm{r} = \sqrt{2\sigma^2 \ln\left(\frac{1}{1-u}\right)}$$

Note that the distribution function $F_R(\mathrm{r}) = 1 - e^{-\frac{r^2}{2\sigma^2}}$ is valid only for positive values of 'r'. It is also noticed that the range for the u is from 0 to 1.

Thus to generate the outcomes with Rayleigh distributed, generate the outcomes 'u', that is uniformly distributed over the range 0–1 and use the formula $\mathrm{r} = \sqrt{2\sigma^2 \ln\left(\frac{1}{1-u}\right)}$ to obtain the Rayleigh distributed outcomes.

Step 2: Generate the sample outcomes of the uniformly distributed random variable 'Θ' that ranges from 0 to 2π.

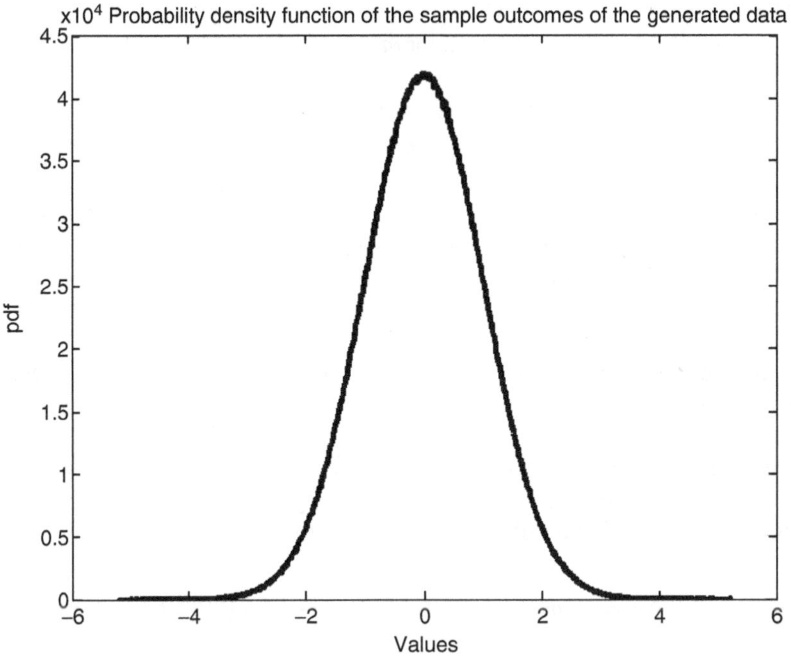

Fig. 6.1 Probability density function of the 10,000,000 samples generated using Matlab (Gaussian distribution with mean = 0 and variance = 1)

Step 3: Compute $X = R cos\Theta$ and $Y = R sin\Theta$ to obtain the two sets of sample outcomes that are independent Gaussian distributed with mean '0' and variance σ^2 (Fig. 6.1).

Steps to Generate the Multivariate (say 'N') Gaussian distributed sample outcomes Repeat the procedure described above to generate 'N' outcomes which are individually Gaussian distributed with mean = 0 and variance = 1. Note that they are independent in nature (i.e.) If we compute the co-variance matrix for the above generated 'N' outcomes, it is identity matrix. Let the generated Multivariate Gaussian distributed outcomes be represented as the outcomes of the random vector 'X'.

Now the requirement is to obtain the outcomes of the Gaussian random vector 'Y' whose mean and co-variance matrix are represented as 'C_Y' and 'M' respectively the generated outcomes of the random vector 'X'.

They are obtained using transformation matrix 'A' as described below.

Let Y = AX + b be the transformation equation which transforms the multivariate random variable X to the multivariate random variable 'Y', where 'A' is the transformation matrix and 'b' is the column vector.

If 'X' is the Multivariate Gaussian distributed, then 'Y' is also the Multivariate Gaussian distributed with Co-variance matrix $C_Y = AC_X\ A^T$ and the mean vector 'Am_X + b', where m_X is the mean vector of the multivariate random variable X.

The generated Multivariate Gaussian distributed outcomes has the Identity co-variance matrix C_X and the mean vector $m_X = 0$,

$$\Rightarrow C_Y = AA^T.$$
$$\Rightarrow \text{Mean vector} = \text{b}$$

Thus the requirement is to represent the required co-variance C_Y matrix as the product of A and A^T and hence the matrix 'A' is obtained. Then applying the transformation $Y = AX + b$ to obtain Multivariate Gaussian distributed outcomes with the specified mean 'b' and the co-variance matrix C_Y. Choose 'b $= m_Y$' for the specifications mentioned above.

Representing the matrix C_Y as the product of AA^T is obtained using Eigen decomposition as described below.

Represent the matrix $C_Y = EDE^T$, where E is the Eigen column matrix, in which the columns are the orthonormal Eigen vectors obtained from the co-variance matrix C_Y and D is the Diagonal matrix with Diagonal elements filled up with the corresponding Eigen values of the co-variance matrix C_Y.

$$C_Y = EDE^T = ED^{\frac{1}{2}} D^{\frac{1}{2}} E^T = \left(ED^{\frac{1}{2}}\right)\left(ED^{\frac{1}{2}}\right)^T.$$
$$\Rightarrow A = \left(ED^{\frac{1}{2}}\right) \text{ is obtained.}$$

Using the obtained transformation matrix 'A', the outcomes of the random vector 'X' is transformed to the outcomes of the random vector 'Y' using the equation $Y = AX + m_Y$.

```
gengd.m
%m-file for generating the Multivariate Gaussian
  distributed
%sample outcomes with zero mean vector and Identity
 matrix
%co-variance matrix
for k = 1:2:10
u = rand(1,1000000);
r = sqrt(2*log(1./(1 - u)));
theta = rand(1,1000000).*2*pi;
Z1 = r.*cos(theta);
Z2 = r.*sin(theta);
X{k} = X;
X{k + 1} = Y;
end
CX = cov(cell2mat(X')');
```

The Co-variance matrix of the generated data X is computed as

$$C_X = \begin{bmatrix} 1.0 & 0 & 0 & 0 & 0 & 0 & 0 & 0 & 0 & 0 \\ 0 & 0.99 & 0 & 0 & 0 & 0 & 0 & 0 & 0 & 0 \\ 0 & 0 & .99 & 0 & 0 & 0 & 0 & 0 & 0 & 0 \\ 0 & 0 & 0 & 1 & 0 & 0 & 0 & 0 & 0 & 0 \\ 0 & 0 & 0 & 0 & .99 & 0 & 0 & 0 & 0 & 0 \\ 0 & 0 & 0 & 0 & 0 & 1.0 & 0 & 0 & 0 & 0 \\ 0 & 0 & 0 & 0 & 0 & 0 & 1.0 & 0 & 0 & 0 \\ 0 & 0 & 0 & 0 & 0 & 0 & 1.0 & 0 & 0 & 0 \\ 0 & 0 & 0 & 0 & 0 & 0 & 0 & 0 & 0.99 & 0 \\ 0 & 0 & 0 & 0 & 0 & 0 & 0 & 0 & 0 & 1. \end{bmatrix}$$

Note that the co-variance matrix of the generated sample outcomes is almost identity matrix. (As expected)

Let mean vector $m_Y = [0.1\ 0.2\ 0.3\ 0.4\ 0.5\ 0.6\ 0.7\ 0.8\ 0.9\ 1.0]$ and

$$C_Y = \begin{bmatrix} 0.1 & 0.2 & 0.3 & 0.4 & 0.5 & 0.6 & 0.7 & 0.8 & 0.9 & 1.0 \\ 0.2 & 0.4 & 0.6 & 0.8 & 1.0 & 1.2 & 1.4 & 1.6 & 1.8 & 2.0 \\ 0.3 & 0.6 & 0.9 & 1.2 & 1.5 & 1.8 & 2.1 & 2.4 & 2.7 & 3.0 \\ 0.4 & 0.8 & 1.2 & 1.6 & 2.0 & 2.4 & 2.8 & 3.2 & 3.6 & 4.0 \\ 0.5 & 1.0 & 1.5 & 2.0 & 2.5 & 3.0 & 3.5 & 4.0 & 4.5 & 5.0 \\ 0.6 & 1.2 & 1.8 & 2.4 & 3.0 & 3.6 & 4.2 & 4.8 & 5.4 & 6.0 \\ 0.7 & 1.4 & 2.1 & 2.8 & 3.5 & 4.2 & 4.9 & 5.6 & 6.3 & 7.0 \\ 0.8 & 1.6 & 2.4 & 3.2 & 4.0 & 4.8 & 5.6 & 6.4 & 7.2 & 8.0 \\ 0.9 & 1.8 & 2.7 & 3.6 & 4.5 & 5.4 & 6.3 & 7.2 & 8.1 & 9.0 \\ 1.0 & 2.0 & 3.0 & 4.0 & 5 & 6.0 & 7.0 & 8.0 & 9.0 & 10.0 \end{bmatrix}$$

```
%m-file for generating the Multivariate Gaussian
 distributed sample
%outcomes with mean vector mY and the co-variance
 matrix Co-
%variance matrix CY as displayed below

MY = 0.1:0.1:1.0;
CY = [1:1:10;2:2:20;3:3:30;4:4:40;5:5:50;6:6:60;7:7:
70;8:8:80;9:9:90;10:10:100]*0.1;
%Representing the matrix CY as the product of two
 matrix A and A' as follows
%As the co-variance matrix is the symmetric matrix,
 the eigen values are
%real
[P,Q] = eig(CY);
A = P*sqrt(abs(Q));
```

```
for i = 1:2:10
r = rand(1,1000000);
sigma = 1;
R = sqrt(2*sigma^2*log(1./(1 - r)));
theta = rand(1,1000000)*2*pi;
u = R.*cos(theta);
v = R.*sin(theta);
[p1,q1] = hist(u,100)
[p2,q2] = hist(v,100)
data{i} = u
data{i+1} = v
end
X = cell2mat(data');
Y = A*X + repmat(MY',1,1000000);
CY = cov(Y');
```

The co-variance matrix computed for the generated Multivariate Gaussian distribution sample outcomes is as shown below (as expected) (Figs. 6.2 and 6.3).

$$C_Y = \begin{bmatrix} 0.1 & 0.2 & 0.3 & 0.4 & 0.5 & 0.6 & 0.7 & 0.8 & 0.9 & 1.0 \\ 0.2 & 0.4 & 0.6 & 0.8 & 1.0 & 1.2 & 1.4 & 1.6 & 1.8 & 2.0 \\ 0.3 & 0.6 & 0.9 & 1.2 & 1.5 & 1.8 & 2.1 & 2.4 & 2.7 & 3.0 \\ 0.4 & 0.8 & 1.2 & 1.6 & 2.0 & 2.4 & 2.8 & 3.2 & 3.6 & 4.0 \\ 0.5 & 1.0 & 1.5 & 2.0 & 2.5 & 3.0 & 3.5 & 4.0 & 4.5 & 5.0 \\ 0.6 & 1.2 & 1.8 & 2.4 & 3.0 & 3.6 & 4.2 & 4.8 & 5.4 & 6.0 \\ 0.7 & 1.4 & 2.1 & 2.8 & 3.5 & 4.2 & 4.9 & 5.6 & 6.3 & 7.0 \\ 0.8 & 1.6 & 2.4 & 3.2 & 4.0 & 4.8 & 5.6 & 6.4 & 7.2 & 8.0 \\ 0.9 & 1.8 & 2.7 & 3.6 & 4.5 & 5.4 & 6.3 & 7.2 & 8.1 & 9.0 \\ 1.0 & 2.0 & 3.0 & 4.0 & 5 & 6.0 & 7.0 & 8.0 & 9.0 & 10.0 \end{bmatrix}$$

6.2 Bacterial Foraging Optimization Technique

Evolutionary algorithms that are formulated from the inspiration of the natural biological behavior are called Biologically Inspired Algorithms (BIA).Bacterial Foraging is one of the BIA inspired from the Foraging behavior of the E-Coli Bacteria. During Foraging, Bacteria tries to move towards the region where more nutrients are available. (i.e.) Moving towards the region where the concentration of nutrients are large. They pass through the neutral medium and they avoid poisonous substances. This biological behavior is inspired to formulate the Bacterial Foraging Optimization technique as described below.

Consider the unconstrained optimization problem of minimizing the function $J(X), X \in \mathbb{R}^m$

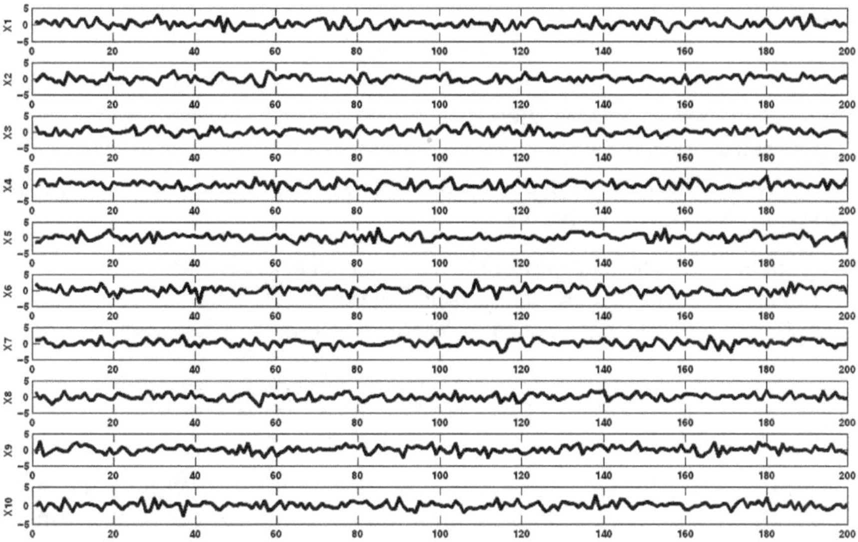

Fig. 6.2 Sample outcomes of the 10-Variate [X1 X2 X3 . . . X10] Gaussian distribution with mean zero and Identity Co-variance matrix

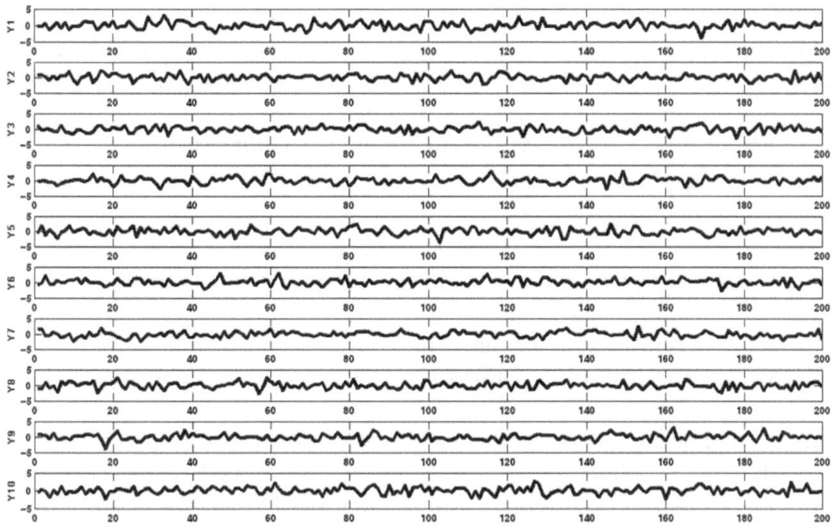

Fig. 6.3 Sample outcomes of the 10-Variate [Y1 Y2 Y3 . . . Y10] Gaussian distribution with mean M_Y and Co-variance matrix C_Y

Analogy: The vector X can be viewed as the position of the Bacteria

$J(X) < 0$ can be treated as the presence of Nutrients
$J(X) = 0$ can be treated as the Neutral
$J(X) > 0$ can be treated as the presence of Toxic substances

Step 1: Initialization of the population
Initialize the positions of 'N' (say) number of Bacteria. Let it be $X^1, X^2, X^3, \ldots X^N$ corresponding to 'N' Bacterium $b^1, b^2, b^3, \ldots b^N$. The position vector $X^1, X^2, X^3, \ldots X^N$ is the initial population.

Step 2: Chemo taxis
It is tendency of the bacteria to move towards the sources of Nutrients. It consists of two stages. They are the following

(a) Tumbling: It is the tendency of the bacteria to change their positions in search of Nutrients. Let X^i_{new} be the next position of the ith Bacteria whose current position is X^i. They are related as described below.

$$X^i_{new} = X^i + c\varphi, \text{ where } \varphi = \frac{\Delta}{\sqrt{\Delta^{\mathrm{T}}\Delta}} \quad \Delta \in \mathbb{R}^m$$

such that each element of the vector Δ is in the range $[-1\ 1]$. Φ is the unit walk in random direction. 'c' is called as chemo tactic step size. The new positions are computed for $i = 1, 2, \ldots N$

(b) Swimming:
Bacterium will tend to keep on moving in the particular direction if it is in the direction that is rich in nutrients.

Mathematically if $J\left(X^i_{new}\right) < J(X^i)$, then another swimming in the same direction (φ) is taken by the ith Bacteria and it can be continued upto N_s steps. After the completion of N_s steps Bacteria goes to the step 3.

If $J\left(X^i_{new}\right) \geq J(X^i)$, Bacteria comes out of the tumbling stage and goes to the step 3.

Step 3: Reproduction
After step 2, best 'N/2' (50%) bacteria measured in terms of its Health are survived. The survived Bacterium are subjected to reproduction to obtain 'N' Bacterium as described below.

Health of the Bacteria is measured in terms of $J(X)$. If the functional value $J(X)$ is less, then the corresponding Bacteria is healthier. Compute $J(X^i)$ *for* $i = 1, 2, \ldots N$. Arrange them in ascending order. First 'N/2' Bacterium and the corresponding positions are selected Let the positions be $[Y^1, Y^2, Y^3, \ldots Y^{N/2}]$. Every Bacteria is split into two Bacterium and are placed in the same positions. Thus new set of positions corresponding to 'N' Bacterium are given as $[Y^1, Y^1, Y^2, Y^2, Y^3, Y^3, \ldots Y^{N/2} Y^{N/2}] = [Z^1, Z^2, Z^3, \ldots Z^N]$ (say). Go to step 2. Repeat 2 and 3 for finite number of iterations. Then Go to step 4.

Step 4: Elimination-dispersion
In real world process, some of the bacterium (i.e.) with probability 'P_d' are dispersed to new locations. This is simulated as shown below.

Generate the random vector of size 1 × N. Sort the elements of the vector in the ascending order. Find the index corresponding to the first N* P_d sorted numbers. Choose the positions of the Bacterium corresponding to the obtained index. They are replaced with the randomly generated positions on the optimization domain.

The positions thus obtained are treated as the current best positions. Go to step 2. Repeat the steps 2–4 for the finite number of iterations.

The best value in every iteration can also be tracked and the best among them can be declared as the optimal solution.

Social Communication

In nature there is the social communication between Bacterium such that they are neither close together nor far away from each other. This is done by releasing the chemical by the Bacteria. The chemical signal can be either attractant or Repellent. If the chemical signal released by the particular Bacteria is attractant in nature, then it attracts other Bacteria to come to its position. On the contrary if the chemical signal released by the particular Bacteria is Repellent in nature, it doesn't allow other Bacteria to come to its position.

The social communication between Bacterium can be simulated using the modified objective function to be computed for the ith position corresponding to the ith position Bacteria as given below.

$$Jmod(X^i) = J(X^i) + Jsocial(X^i), \ where$$

Jmod is the modified Objective function computed for the ith position X^i corresponding to the ith Bacteria. $J(X^i)$ is the actual objective function value computed for the ith position X^i corresponding to the ith Bacteria. $social(X^i)$ is the attractant cum repellent signal computed for the ith position X^i corresponding to the ith Bacteria as displayed below.

$$Let\ d_{ij} = \left\| X^i - X^j \right\|^2$$

$$Jsocial(X^i) = M\left(\sum_{j=1}^{N} e^{-Rd_{ij}} - \sum_{j=1}^{N} e^{-Ad_{ij}}\right)$$

Note that if the first term $\sum_{j=1}^{N} e^{-Rd_{ij}}$ is reduced if distance between the ith position and others are made large and hence it acts as the repellent signal. Similarly the second term $-\sum_{j=1}^{N} e^{-Rd_{ij}}$ is reduced if the distance between the ith position and others are made small and hence it acts as the attractant signal. 'R' is the Repellent factor and 'A' is the attractant factor.

The convergence graph obtained using Bacterial Foraging technique for minimizing the function $J(X) = (x1^\wedge2 - 9)^\wedge2 + (x2^\wedge2 - 9)^\wedge2 + (x3^\wedge2 - 16)^\wedge2$ is shown in the figure given below (Fig. 6.4). The corresponding Matlab program is also displayed below

After 100 iteration, the best solution obtained is [2.9025 2.9512 4.0720]

```
bactalgo.m
%Bacterial Foraging Technique
%Edit bactfun.m to insert the objective function to
 be minimized
%TRACEERROR traces the minimum error obtained in
 every iteration
```

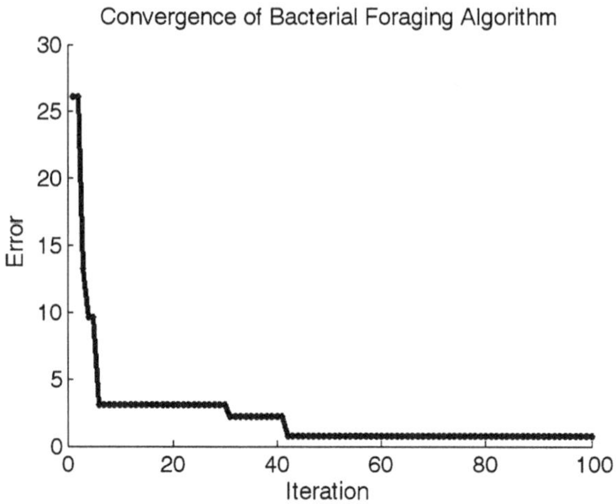

Fig. 6.4 Illustration of the convergence of the bacterial foraging algorithm

```
%TRACEVAL traces the corresponding best solution in
 every iteration
%jsocial regulates the social communication between
 bacterium.
pos1 = rand(1,100)*10;
pos2 = rand(1,100)*10;
pos3 = rand(1,100)*10;
vect = [pos1;pos2;pos3]';
figure
TRACEVALUE = [];
TRACEVECTOR = [];
c = 1;
for i = 1:1:100
  Jcurvalue(i) = bactfun(vect(i,:));
  Jswcurvalue(i) = Jcurvalue(i) + jsocial(vect(i,:),
  vect);
End
%Iteration starts
for iter = 1:1:100
for dispersal = 1:1:50
for survey = 1:1:50
for bact = 1:1:100
  pos1 = rand(1,1)*10;
  pos2 = rand(1,1)*10;
  pos3 = rand(1,1)*10;
  phi = [pos1 pos2 pos3];
  newvect = vect(bact,:) + c*phi;
```

```
    Jcurvaluenew(bact) = bactfun(newvect);
    Jswcurvaluenew(bact) = Jcurvalue(bact) + jsocial
    (newvect,vect);
    %m is the maximum number swimming
    for m = 1:1:25
      if (Jswcurvaluenew(bact) < Jswcurvalue(bact))
        vect(bact,:) = newvect;
        Jcurvalue(bact) = bactfun(vect(bact,:));
        Jswcurvalue(bact) = Jcurvalue(bact) + jsocial
        (vect(bact,:),vect);
        newvect = vect(bact,:) + c*phi;
        Jcurvaluenew(bact) = bactfun(newvect);
        Jswcurvaluenew(bact) = Jswcurvaluenew(bact) +
        jsocial(newvect,vect);
      else
        m = 25;
      end
    end
  end
  %Reproduction of bacteria
  [p,q] = sort(Jswcurvaluenew);
  vect1 = vect(q(1:1:50),:);
  vect = [vect1;vect1];
  end
  %Dispersal of bacteria with probability 0.2
  [p,q] = sort(rand(1,100));
  tempvect1 = vect(q(1:1:80),:);
  pos1 = rand(1,20)*10;
  pos2 = rand(1,20)*10;
  pos3 = rand(1,20)*10;
  tempvect2 = [pos1;pos2;pos3]';
  vect = [tempvect1;tempvect2];
  for i = 1:1:100
    Jcurvalue(i) = bactfun(vect(i,:));
    Jswcurvalue(i) = Jcurvalue(i) + jsocial(vect(i,:),
    vect);
  end
  end
  [p,q] = sort(Jswcurvalue);
  temp = vect(q(1),:);
  TRACEVALUE = [TRACEVALUE bactfun(temp)];
  TRACEVECTOR = [TRACEVECTOR; temp];
  hold on
  plot(TRACEVALUE)
  pause(0.2);
```

```
end
%Solution
for i = 1:1:100
  Jcurvalue(i) = bactfun(vect(i,:));
end
[p,q] = sort(TRACEVALUE);
FINALRES = TRACEVECTOR(q(1),:);
jsocial.m
function [res] = jsocial(a,vect)
M = 1;
b = repmat(a,[100 1]);
vect = vect - b;
vect = vect.^2;
vect = sum(vect');
Wa = 2;
Wr = 2;
res1 = sum(exp(vect*(-Wa)));
res2 = sum(exp(vect*(-Wr)));
bactfun.m
function [res] = bactfun(z);
p = z(1);
q = z(2);
r = z(3);
res = (p^2 - 9)^2 + (q^2 - 9)^2 + (r^2 - 16)^2;
```

6.3 Particle Swarm Optimization

The Particle swarm optimization is the biologically inspired algorithm inspired from the behavior of birds on deciding the optimal path to move from the particular source to the destination. Consider the task of movement of group of birds (say A,B,C) from the particular source point 'S' to the destination 'D'. Let 'A' (which is currently at point P_A) the decides to move towards the point P'_A. Similarly the bird 'B' and 'C' decides to move towards the point P'_B and P'_C from P_B and P_C respectively. Let the distance between the point P'_C and the destination point 'D' is less compared to the distance between the point P'_B and the destination point 'D' and the distance between the point P'_B and the destination point 'D'. Thus the final decision taken by the bird 'B' is the combination of the individual decision taken by bird 'B' and the best global decision taken by the neighboring birds (In the current scenario, the best global decision is the decision taken by the bird 'C').

Mathematically the bird moves from the current position to the next position described as follows.

The next position moved by the bird 'B' is

$$P_B + g_B\left(P'_C - P_B\right) + l_B\left(P'_B - P_B\right)$$

Similarly the next position moved by the bird 'A' and 'C' are given below

$$P_A + g_A\left(P'_C - P_A\right) + l_A\left(P'_B - P_A\right)$$
$$P_C + g_C\left(P'_C - P_C\right)$$

where 'g_A, g_B, g_C' are the global constants and 'l_A, l_B' are the local constants.

Consider the unconstrained optimization problem of minimizing the function $J(X), X \in \mathbb{R}^m$.

Analogy:

The vector 'X' in the above definition is treated as the current position of the bird in the PSO algorithm. The corresponding value $J(X)$ is the distance between the current position of the bird and the destination. The PSO algorithm identifies the shortest path as described above so that it reaches the destination as early as possible. (i.e.) Identifying the optimal value of X such that $J(X)$ is minimized

Algorithm

Step 1: Initialize the positions of 'N' number of birds. Let it be $X_1, X_2, X_3, \ldots X_N$.
Step 2: Obtain the next positions of the 'N' birds using the combination of local decision taken by the individual birds and the best decision taken by the neighboring birds (as described earlier).
Step 3: The next positions thus obtained are treated as current positions.

Local decisions taken by the individual birds for the next move are taken as per the procedure given below.

If the distance between current local decision taken by the ith bird position and the destination is greater than the distance between the next position mentioned in the step 2 and the destination, next position is treated as the local decision taken by the ith bird position for the next movement. Otherwise the current local decision taken by the ith bird position is considered as the local decision for the next move also. Repeat step 2 and 3 for the finite number of iterations.
Step 4: Best position among the 'N' positions of the birds in the last iteration corresponding to 'N' birds is declared as the final solution for minimizing the function $J(X)$. Best position among the 'N' positions in the final iteration is the position whose distance from the destination is minimum among all other positions.

(i.e.) Best position$= arg_i(\|X_i{-}D\|), i = 1, 2, 3, \ldots N$

The convergence graph obtained using Particle Swarm Optimization for minimizing the function $J(X) = 2(x1-4)\text{^}2 + 4(x2-3)\text{^}2 + 5(x3-6)\text{^}2$ is shown in the Fig. 6.5. The corresponding Matlab program is also displayed below.

For the problem mentioned above the vector of the form $\begin{bmatrix} x1 \\ x2 \\ x3 \end{bmatrix}$ is treated as the arbitrary position of the bird. PSO algorithm is performed to obtain the optimal value of the position of the bird such that its distance from the destination is minimized. The distance between the current position of the bird and the destination is measured using the objective function $J(X)$ and hence the function is minimized.

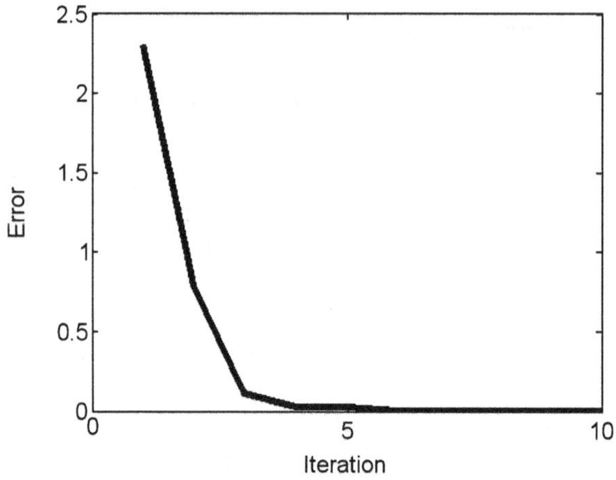

Fig. 6.5 Illustration of the convergence of the PSO algorithm

Immediate after fifth iteration, the solution obtained using PSO algorithm is given as [3.9999 3 6] and the is corresponding

6.4 Newton's Iterative Method

Consider the problem of minimizing the function $f(x1, x2, x3) = 2(x1-4)^2 + 4(x2-3)^2 + 5(x3-6)^2$

Expressing the above function using Taylor series we get the following.

$$f\left(\begin{bmatrix} x1 \\ x2 \\ x3 \end{bmatrix} + \begin{bmatrix} \Delta x1 \\ \Delta x2 \\ \Delta x3 \end{bmatrix}\right) = f\left(\begin{bmatrix} x1 \\ x2 \\ x3 \end{bmatrix}\right) + Df\left(\begin{bmatrix} x1 \\ x2 \\ x3 \end{bmatrix}\right) + \frac{D^2}{2!} f^2\left(\begin{bmatrix} x1 \\ x2 \\ x3 \end{bmatrix}\right) + \ldots$$

where D $= \Delta x1 \frac{\partial}{\partial x1} + \Delta x2 \frac{\partial}{\partial x2} + \Delta x3 \frac{\partial}{\partial x3}$.

The above series can also be represented in the matrix form as shown below

$$f\left(\begin{bmatrix} x1 \\ x2 \\ x3 \end{bmatrix} + \begin{bmatrix} \Delta x1 \\ \Delta x2 \\ \Delta x3 \end{bmatrix}\right)$$

$$= f\left(\begin{bmatrix} x1 \\ x2 \\ x3 \end{bmatrix}\right) + [\Delta x1 \ \Delta x2 \ \Delta 3] \begin{bmatrix} \frac{\partial f}{\partial x1} \\ \frac{\partial f}{\partial x1} \\ \frac{\partial f}{\partial x1} \end{bmatrix}$$

$$+ [\Delta x1\ \Delta x2\ \Delta 3] \begin{bmatrix} \frac{\partial^2 f}{\partial x1^2} & \frac{\partial^2 f}{\partial x1 \partial x2} & \frac{\partial^2 f}{\partial x1 \partial x3} \\ \frac{\partial^2 f}{\partial x2 \partial x1} & \frac{\partial^2 f}{\partial x2^2} & \frac{\partial^2 f}{\partial x2 \partial x3} \\ \frac{\partial^2 f}{\partial x3 \partial x1} & \frac{\partial^2 f}{\partial x3 \partial x2} & \frac{\partial^2 f}{\partial x3^2} \end{bmatrix} \begin{bmatrix} \Delta x1 \\ \Delta x2 \\ \Delta x3 \end{bmatrix}$$

$$+ \ldots$$

Let $\begin{bmatrix} x1 \\ x2 \\ x3 \end{bmatrix} + \begin{bmatrix} \Delta x1 \\ \Delta x2 \\ \Delta x3 \end{bmatrix} = \begin{bmatrix} x_1^{n+1} \\ x_2^{n+1} \\ x_3^{n+1} \end{bmatrix}$ and $\begin{bmatrix} x1 \\ x2 \\ x3 \end{bmatrix} = \begin{bmatrix} x_1^{n} \\ x_2^{n} \\ x_3^{n} \end{bmatrix}$

Rewriting the Taylor series using the notations used above we get,

$$f\left(\begin{bmatrix} x_1^{n+1} \\ x_2^{n+1} \\ x_3^{n+1} \end{bmatrix}\right) = f\left(\begin{bmatrix} x_1^{n} \\ x_2^{n} \\ x_3^{n} \end{bmatrix}\right) + \left[x_1^{n+1} - x_1^n \quad x_2^{n+1} - x_2^n \quad x_3^{n+1} - x_3^n\right]$$

$$\begin{bmatrix} \frac{\partial f}{\partial x_1} \\ \frac{\partial f}{\partial x_2} \\ \frac{\partial f}{\partial x_3} \end{bmatrix} \left(at \begin{bmatrix} x_1^{n} \\ x_2^{n} \\ x_3^{n} \end{bmatrix}\right) + \left[x_1^{n+1} - x_1^n \quad x_2^{n+1} - x_2^n \quad x_3^{n+1} - x_3^n\right]$$

$$\begin{bmatrix} \frac{\partial^2 f}{\partial x_1 \partial x_1} & \frac{\partial^2 f}{\partial x_1 \partial x_2} & \frac{\partial^2 f}{\partial x_1 \partial x_3} \\ \frac{\partial^2 f}{\partial x_2 \partial x_1} & \frac{\partial^2 f}{\partial x_2 \partial x_2} & \frac{\partial^2 f}{\partial x_2 \partial x_3} \\ \frac{\partial^2 f}{\partial x_3 \partial x_1} & \frac{\partial^2 f}{\partial x_3 \partial x_2} & \frac{\partial^2 f}{\partial x_3 \partial x_3} \end{bmatrix} \left(at \begin{bmatrix} x_1^{n} \\ x_2^{n} \\ x_3^{n} \end{bmatrix}\right) \begin{bmatrix} x_1^{n+1} - x_1^n \\ x_2^{n+1} - x_2^n \\ x_3^{n+1} - x_3^n \end{bmatrix} + \ldots$$

If we want to find out the roots of the above equation, we equate $f\left(\begin{bmatrix} x_1^{n+1} \\ x_2^{n+1} \\ x_3^{n+1} \end{bmatrix}\right) = 0$

and solve for the vector $\begin{bmatrix} x_1^{n+1} \\ x_2^{n+1} \\ x_3^{n+1} \end{bmatrix}$ as shown below

(Considering only first two terms)

$$f\left(\begin{bmatrix} x_1^n \\ x_2^n \\ x_3^n \end{bmatrix}\right) = -\begin{bmatrix} x_1^{n+1}-x_1^n & x_2^{n+1}-x_2^n & x_3^{n+1}-x_3^n \end{bmatrix}$$

$$\begin{bmatrix} \frac{\partial f}{\partial x_1} \\ \frac{\partial f}{\partial x_2} \\ \frac{\partial f}{\partial x_3} \end{bmatrix}\left(at \begin{bmatrix} x_1^n \\ x_2^n \\ x_3^n \end{bmatrix}\right)$$

$$\Rightarrow \begin{bmatrix} x_1^{n+1} & x_2^{n+1} & x_2^{n+1} \end{bmatrix} = \begin{bmatrix} x_1^n & x_2^n & x_2^n \end{bmatrix} - f\left(\begin{bmatrix} x_1^n \\ x_2^n \\ x_3^n \end{bmatrix}\right)\begin{bmatrix} \frac{\partial f}{\partial x_1} \\ \frac{\partial f}{\partial x_2} \\ \frac{\partial f}{\partial x_3} \end{bmatrix}^{-1}\left(at \begin{bmatrix} x_1^n \\ x_2^n \\ x_3^n \end{bmatrix}\right)$$

The above mentioned equation is called Newton's method of computing the roots of the multivariable equation $f(x1, x2, x3)$.

Note:

Note that $\begin{bmatrix} \frac{\partial f}{\partial x_1} \\ \frac{\partial f}{\partial x_2} \\ \frac{\partial f}{\partial x_3} \end{bmatrix}\left(at \begin{bmatrix} x_1^n \\ x_2^n \\ x_3^n \end{bmatrix}\right)$ is the column vector and hence the inverse mentioned in the equation is the pseudo inverse (see Chapter 1 for details).

Consider the Taylor series equation as mentioned below

$$f\left(\begin{bmatrix} x_1^{n+1} \\ x_2^{n+1} \\ x_3^{n+1} \end{bmatrix}\right) = f\left(\begin{bmatrix} x_1^n \\ x_2^n \\ x_3^n \end{bmatrix}\right) + \begin{bmatrix} x_1^{n+1}-x_1^n & x_2^{n+1}-x_2^n & x_3^{n+1}-x_3^n \end{bmatrix}$$

$$\begin{bmatrix} \frac{\partial f}{\partial x_1} \\ \frac{\partial f}{\partial x_2} \\ \frac{\partial f}{\partial x_3} \end{bmatrix}\left(at \begin{bmatrix} x_1^n \\ x_2^n \\ x_3^n \end{bmatrix}\right) + \ldots$$

Differentiating with respect to the vector $\begin{bmatrix} x_1 \\ x_2 \\ x_3 \end{bmatrix}$ on both sides we get the following

$$\nabla \mathrm{f}\left(\mathrm{at}\begin{bmatrix} x_1^{n+1} \\ x_2^{n+1} \\ x_3^{n+1} \end{bmatrix}\right) = \begin{bmatrix} \frac{\partial f}{\partial x1} \\ \frac{\partial f}{\partial x2} \\ \frac{\partial f}{\partial x3} \end{bmatrix}\left(at\begin{bmatrix} x_1^{n} \\ x_2^{n} \\ x_3^{n} \end{bmatrix}\right) +$$

$$\begin{bmatrix} \frac{\partial^2 f}{\partial x_1 \partial x_1} & \frac{\partial^2 f}{\partial x_1 \partial x_2} & \frac{\partial^2 f}{\partial x_1 \partial x_3} \\ \frac{\partial^2 f}{\partial x_2 \partial x_1} & \frac{\partial^2 f}{\partial x_2 \partial x_2} & \frac{\partial^2 f}{\partial x_2 \partial x_3} \\ \frac{\partial^2 f}{\partial x_3 \partial x_1} & \frac{\partial^2 f}{\partial x_3 \partial x_2} & \frac{\partial^2 f}{\partial x_3 \partial x_3} \end{bmatrix}\left(at\begin{bmatrix} x_1^{n} \\ x_2^{n} \\ x_3^{n} \end{bmatrix}\right)\begin{bmatrix} x_1^{n+1} - x_1^{n} \\ x_2^{n+1} - x_2^{n} \\ x_3^{n+1} - x_3^{n} \end{bmatrix}$$

Also to consider the point $\begin{bmatrix} x_1^{n+1} \\ x_2^{n+1} \\ x_3^{n+1} \end{bmatrix}$ as the extremal point of the equation $f(x1, x2, x3)$, then $\nabla \mathrm{f}\left(\mathrm{at}\begin{bmatrix} x_1^{n+1} \\ x_2^{n+1} \\ x_3^{n+1} \end{bmatrix}\right) = 0.$

Using the above condition and solving the expression for extremal point we get

$$\begin{bmatrix} x_1^{n+1} \\ x_2^{n+1} \\ x_3^{n+1} \end{bmatrix} =$$

$$\begin{bmatrix} x_1^{n} \\ x_2^{n} \\ x_3^{n} \end{bmatrix} - \begin{bmatrix} \frac{\partial^2 f}{\partial x_1 \partial x_1} & \frac{\partial^2 f}{\partial x_1 \partial x_2} & \frac{\partial^2 f}{\partial x_1 \partial x_3} \\ \frac{\partial^2 f}{\partial x_2 \partial x_1} & \frac{\partial^2 f}{\partial x_2 \partial x_2} & \frac{\partial^2 f}{\partial x_2 \partial x_3} \\ \frac{\partial^2 f}{\partial x_3 \partial x_1} & \frac{\partial^2 f}{\partial x_3 \partial x_2} & \frac{\partial^2 f}{\partial x_3 \partial x_3} \end{bmatrix}^{-1}\left(at\begin{bmatrix} x_1^{n} \\ x_2^{n} \\ x_3^{n} \end{bmatrix}\right)\begin{bmatrix} \frac{\partial f}{\partial x1} \\ \frac{\partial f}{\partial x2} \\ \frac{\partial f}{\partial x3} \end{bmatrix}\left(at\begin{bmatrix} x_1^{n} \\ x_2^{n} \\ x_3^{n} \end{bmatrix}\right)$$

Consider the problem of minimizing the function $f(x1, x2, x3) = 2(x1 - 4)^2 + 4(x2 - 3)^2 + 5(x3 - 6)^2$

Newton's iteration equation is formulated as shown below.

$$\begin{bmatrix} x_1^{n+1} \\ x_2^{n+1} \\ x_3^{n+1} \end{bmatrix} = \begin{bmatrix} x_1^{n} \\ x_2^{n} \\ x_3^{n} \end{bmatrix} - \begin{bmatrix} 1/4 & 0 & 0 \\ 0 & 1/8 & 0 \\ 0 & 0 & 1/10 \end{bmatrix} \begin{bmatrix} 4\left(x_1^n - 4\right) \\ 8\left(x_1^n - 3\right) \\ 12\left(x_3^n - 6\right) \end{bmatrix}$$

Let us initialize the extremum vector as $\begin{bmatrix} 0 \\ 0 \\ 0 \end{bmatrix}$

Iterations are performed using Matlab and Error (vs) Iteration table is mentioned below for illustration.

Iteration	1	2	3	4	5
Error	7.2	0.2880	0.0115	0.0005	0.0000

Note that Error function converges to zero immediate after reaching fifth iteration.

6.5 Steepest Descent Algorithm

Consider the problem of minimizing the function $f(x1, x2, x3) = 2(x1 - 4)^2 + 4(x2 - 3)^2 + 5(x3 - 6)^2$

The Linear approximation of the curve $f(x1, x2, x3)$ can be obtained as follows

$$f(x1 + k\,\lambda_1, x2 + k\,\lambda_2, x3 + k\,\lambda_3)$$
$$= f(x1, x2, x3) + k \begin{bmatrix} \frac{\partial f}{\partial x1} & \frac{\partial f}{\partial x2} & \frac{\partial f}{\partial x3} \end{bmatrix} \begin{bmatrix} \lambda_1 \\ \lambda_2 \\ \lambda_3 \end{bmatrix}$$
$$(i.e) f(X + k\lambda) = f(X) + k\nabla f^T \lambda$$

$\begin{bmatrix} \lambda_1 \\ \lambda_2 \\ \lambda_3 \end{bmatrix}$ is the unit vector and k is some scalar constant

The maximum increase in the value of the function $f(x1, x2, x3)$ (i.e.) $f(x1 + k\,\lambda_1, x2 + k\lambda_2, x3 + k\lambda_3) - f(x1, x2, x3)$ occurs when $\begin{bmatrix} \frac{\partial f}{\partial x1} & \frac{\partial f}{\partial x2} & \frac{\partial f}{\partial x3} \end{bmatrix} \begin{bmatrix} \lambda_1 \\ \lambda_2 \\ \lambda_3 \end{bmatrix}$ is maximum. We also know from Cauchy-Schwarz inequality (see Chapter 4) that maximum value of the inner product $\begin{bmatrix} \frac{\partial f}{\partial x1} & \frac{\partial f}{\partial x2} & \frac{\partial f}{\partial x3} \end{bmatrix} \begin{bmatrix} \lambda_1 \\ \lambda_2 \\ \lambda_3 \end{bmatrix}$ occurs only when

$\begin{bmatrix} \lambda_1 \\ \lambda_2 \\ \lambda_3 \end{bmatrix} = l \begin{bmatrix} \frac{\partial f}{\partial x1} \\ \frac{\partial f}{\partial x2} \\ \frac{\partial f}{\partial x3} \end{bmatrix}$, where 'l' is some scalar constant.

Similarly maximum decrease in the function occurs when $\begin{bmatrix} \lambda_1 \\ \lambda_2 \\ \lambda_3 \end{bmatrix} = -l \begin{bmatrix} \frac{\partial f}{\partial x1} \\ \frac{\partial f}{\partial x2} \\ \frac{\partial f}{\partial x3} \end{bmatrix}$

Let $\begin{bmatrix} x_1^n \\ x_2^n \\ x_2^n \end{bmatrix}$ be the current value of the vector X used in the steepest descent iteration algorithm. Then the next best value for the vector X represented as $\begin{bmatrix} x_1^{n+1} \\ x_2^{n+1} \\ x_2^{n+1} \end{bmatrix}$ such that the vector is in the direction of decreasing function $f(\mathrm{X})$ is given as follows.

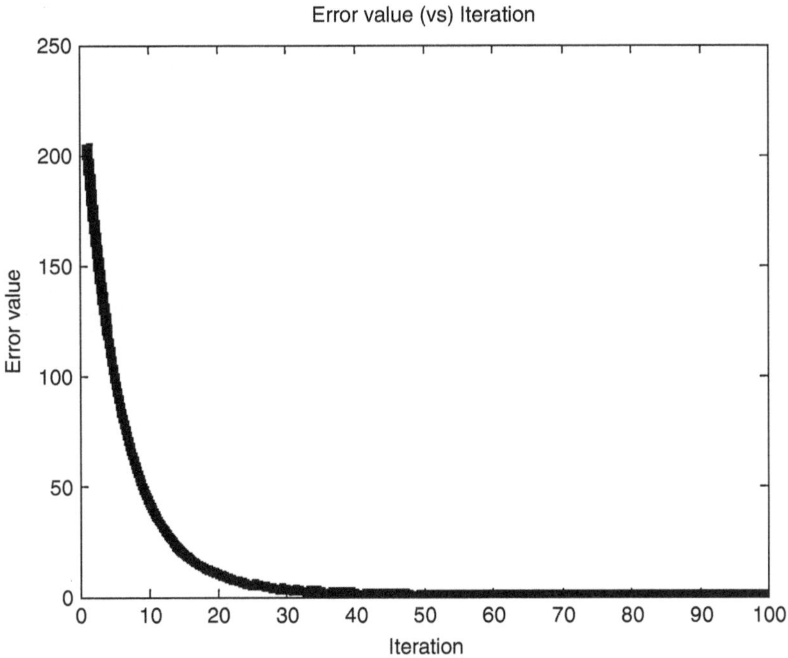

Fig. 6.6 Illustration of the convergence of the steepest descent algorithm

$$\begin{bmatrix} x_1^{n+1} \\ x_2^{n+1} \\ x_2^{n+1} \end{bmatrix} = \begin{bmatrix} x_1^n \\ x_2^n \\ x_2^n \end{bmatrix} - l \begin{bmatrix} \frac{\partial f}{\partial x1} \\ \frac{\partial f}{\partial x2} \\ \frac{\partial f}{\partial x3} \end{bmatrix}$$

For the above problem the iterative equation is obtained as follows

$$\begin{bmatrix} x_1^{n+1} \\ x_2^{n+1} \\ x_2^{n+1} \end{bmatrix} = \begin{bmatrix} x_1^n \\ x_2^n \\ x_2^n \end{bmatrix} - l \begin{bmatrix} 4\,(x1-4) \\ 8\,(x2-3) \\ 10\,(x3-6) \end{bmatrix}$$

Let us initialize the vector $= \begin{bmatrix} 0 \\ 0 \\ 0 \end{bmatrix}$. Iterations are performed using Matlab and the convergence graph is plotted for illustration (Fig. 6.6).

The learning rate is chosen as $l = 0.01$. After 100 iterations, function value reaches 0.0091 and the corresponding vector obtained is $\begin{bmatrix} 3.9325 \\ 2.9993 \\ 5.9998 \end{bmatrix}$.

Index

FSC
www.fsc.org
MIX
Papier | Fördert gute Waldnutzung
FSC® C083411

Kolibri 360 GmbH
Ferdinand-Jühlke-Straße 7
99095 Erfurt, Deutschland
produktsicherheit@kolibri360.de